NUCLEAR POWER

The History of Nuclear Power

JAMES A. MAHAFFEY, PH.D.

Facts On File
An Infobase Learning Company

For Katherine Grace Whatley

The History of Nuclear Power

Copyright © 2011 by James A. Mahaffey, Ph.D.

Facts On File, Inc.
An imprint of Infobase Learning
132 West 31st Street
New York NY 10001

Library of Congress Cataloging-in-Publication Data
Mahaffey, James A.
  The history of nuclear power / [by James A. Mahaffey].
    p. cm.—(Nuclear power)
  Includes bibliographical references and index.
  ISBN 978-0-8160-7649-9
1. Nuclear energy—History—Popular works. I. Title.
  QC773.M26 2011
  333.792′409—dc22   2010043236

Facts On File books are available at special discounts when purchased in bulk quantities for businesses, associations, institutions, or sales promotions. Please call our Special Sales Department in New York at (212) 967-8800 or (800) 322-8755.

You can find Facts On File on the World Wide Web at http://www.infobaselearning.com

Excerpts included herewith have been reprinted by permission of the copyright holders; the author has made every effort to contact copyright holders. The publishers will be glad to rectify, in future editions, any errors or omissions brought to their notice.

Text design and composition by by Annie O'Donnell
Illustrations by Bobbi McCutcheon
Photo research by Suzanne M. Tibor
Cover printed by Yurchak Printing, Landisville, Pa.
Book printed and bound by Yurchak Printing, Landisville, Pa.
Date printed: August 2011
Printed in the United States of America

10 9 8 7 6 5 4 3 2 1

This book is printed on acid-free paper.

Contents

Preface   vii

Acknowledgments   x

Introduction   xi

1 Centuries of Atomic Structure Theories   1

Earliest Concepts of Atomic Structure   2

Fluorescence and the Discovery of Radioactivity   5

   Evidence of Prehistoric Nuclear Activity   6

Proof That Atoms Can Be Broken   12

Marie and Pierre Curie Find Radium in Uranium Ore   15

2 Discovery of the Atomic Nucleus   18

Ernest Rutherford Starts Naming Rays and Particles   19

The Energy Released by Radioactive Decay   21

   Ernest Rutherford: The Man Who Sorted Out the Atomic Structure   22

The Discovery of the Atomic Nucleus   23

3 Monumental Theories   27

Max Planck (1858–1947) and the Elementary Quantum of Action   28

The Copenhagen Interpretation   30

Niels Bohr Imposes Quantum Mechanics on the Atomic Model   32

   Counterintuitive Aspects of Quantum Mechanics   32

4 Nuclear Fission Is Discovered   35

James Chadwick Proposes a Neutron   36

Leó Szilárd Thinks of the Self-Sustaining Chain Reaction   38

The Remote Collaboration of Otto Hahn (1879–1968) and

   Lise Meitner (1878–1968)   41

   Lise Meitner: A Refugee Scientist   44

The Race Is On   44

5   A Gathering of Nuclear Scientists in the United States   47

The Feared Threat of a German Atomic Bomb   48

The Interesting Effects of Neutrons at Low Speeds   50

    Niels Bohr: The Last of the Refugees   52

An Exodus from Europe   55

Preliminary Nuclear Research in the United States   56

6   The First Sustained Nuclear Power Production   59

A Letter to the President of the United States from Albert Einstein   60

    The Need for Secrecy   63

The First Nuclear Reactor   64

The Manhattan Project Begins   70

7   Nuclear Weaponry Development   73

First Work at the Los Alamos Laboratory   74

Two Atomic Bomb Designs Diverge   80

    Espionage in the Laboratory   81

Nuclear Weapons Research in Germany, Japan, and
    the Soviet Union   83

The Trinity Test   86

Japan Surrenders   90

8   Atoms for Peace and Atoms for War   93

The Building of the *Nautilus*   94

The Atomic Energy Act and Atoms for Peace   101

    Admiral Hyman Rickover: Father of the Nuclear Navy   102

The BORAX Reactors in Idaho   106

9   America Goes Nuclear   109

The First Civilian Power Reactors   111

    Safety Analysis   116

Nuclear Power Becomes Commercial   116

The Environmental Protection Agency and Long-Term
    Spent-Fuel Storage   124

Nuclear Power Goes into a Long Sleep   128

New Realities   129

Conclusion 132

Chronology 134

Glossary 142

Further Resources 149

Index 157

Nuclear Power is a multivolume set that explores the inner workings, history, science, global politics, future hopes, triumphs, and disasters of an industry that was, in a sense, born backward. Nuclear technology may be unique among the great technical achievements, in that its greatest moments of discovery and advancement were kept hidden from all except those most closely involved in the complex and sophisticated experimental work related to it. The public first became aware of nuclear energy at the end of World War II, when the United States brought the hostilities in the Pacific to an abrupt end by destroying two Japanese cities with atomic weapons. This was a practical demonstration of a newly developed source of intensely concentrated power. To have wiped out two cities with only two bombs was unique in human experience. The entire world was stunned by the implications, and the specter of nuclear annihilation has not entirely subsided in the 60 years since Hiroshima and Nagasaki.

The introduction of nuclear power was unusual in that it began with specialized explosives rather than small demonstrations of electrical-generating plants, for example. In any similar industry, this new, intriguing source of potential power would have been developed in academic and then industrial laboratories, first as a series of theories, then incremental experiments, graduating to small-scale demonstrations, and, finally, with financial support from some forward-looking industrial firms, an advantageous, alternate form of energy production having an established place in the industrial world. This was not the case for the nuclear industry. The relevant theories required too much effort in an area that was too risky for the usual industrial investment, and the full engagement and commitment of governments was necessary, with military implications for all developments. The future, which could be accurately predicted to involve nuclear power, arrived too soon, before humankind was convinced that renewable energy was needed. After many thousands of years of burning things as fuel, it was a hard habit to shake. Nuclear technology was never developed with public participation, and the atmosphere of secrecy and danger surrounding it eventually led to distrust and distortion. The nuclear power industry exists today, benefiting civilization with a respectable percentage of the total energy supply, despite the unusual lack of understanding and general knowledge among people who tap into it.

This set is designed to address the problems of public perception of nuclear power and to instill interest and arouse curiosity for this branch of technology. *The History of Nuclear Power,* the first volume in the set, explains how a full understanding of matter and energy developed as science emerged and developed. It was only logical that eventually an atomic theory of matter would emerge, and from that a nuclear theory of atoms would be elucidated. Once matter was understood, it was discovered that it could be destroyed and converted directly into energy. From there it was a downhill struggle to capture the energy and direct it to useful purposes.

Nuclear Accidents and Disasters, the second book in the set, concerns the long period of lessons learned in the emergent nuclear industry. It was a new way of doing things, and a great deal of learning by accident analysis was inevitable. These lessons were expensive but well learned, and the body of knowledge gained now results in one of the safest industries on Earth. *Radiation,* the third volume in the set, covers radiation, its long-term and short-term effects, and the ways that humankind is affected by and protected from it. One of the great public concerns about nuclear power is the collateral effect of radiation, and full knowledge of this will be essential for living in a world powered by nuclear means.

Nuclear Fission Reactors, the fourth book in this set, gives a detailed examination of a typical nuclear power plant of the type that now provides 20 percent of the electrical energy in the United States. *Fusion,* the fifth book, covers nuclear fusion, the power source of the universe. Fusion is often overlooked in discussions of nuclear power, but it has great potential as a long-term source of electrical energy. *The Future of Nuclear Power,* the final book in the set, surveys all that is possible in the world of nuclear technology, from spaceflights beyond the solar system to power systems that have the potential to light the Earth after the Sun has burned out.

At the Georgia Institute of Technology, I earned a bachelor of science degree in physics, a master of science, and a doctorate in nuclear engineering. I remained there for more than 30 years, gaining experience in scientific and engineering research in many fields of technology, including nuclear power. Sitting at the control console of a nuclear reactor, I have cold-started the fission process many times, run the reactor at power, and shut it down. Once, I stood atop a reactor core. I also stood on the bottom core plate of a reactor in construction, and on occasion I watched the eerie blue glow at the heart of a reactor running at full power. I did some time in a radiation suit, waved the Geiger counter probe, and spent many days and nights counting neutrons. As a student of nuclear technology, I bring a near-complete view of this, from theories to daily operation of a power plant. Notes and apparatus from my nuclear fusion research have been requested by and given to the National Museum of American History of the Smithsonian Institution. My friends, superiors, and competitors for research funds were people who served on the USS *Nautilus* nuclear submarine, those who assembled the early atomic bombs, and those who were there when nuclear power was born. I knew to listen to their tales.

The Nuclear Power set is written for those who are facing a growing world population with fewer resources and an increasingly fragile environment. A deep understanding of physics, mathematics, or the specialized vocabulary of nuclear technology is not necessary to read the books in this series and grasp what is going on in this important branch of science. It is hoped that you can understand the problems, meet the challenges, and be ready for the future with the information in these books. Each volume in the set includes an index, a chronology of important events, and a glossary of scientific terms. A list of books and Internet resources for further information provides the young reader with additional means to investigate every topic, as the study of nuclear technology expands to touch every aspect of the technical world.

Acknowledgments

I wish to thank Dr. Don S. Harmer, retired Professor Emeritus from the Georgia Institute of Technology School of Physics, an old friend from the Old School who not only taught me much of what I know in the field of nuclear physics but also did a thorough and constructive technical edit of the manuscript. Thanks also to Dr. Douglas E. Wrege, a physicist, a teacher, and a friend from Georgia Tech, who also read the manuscript, finding exotic errors that apparently only he could detect. Special credit is due Frank K. Darmstadt, my editor at Facts On File, who helped me at every step in making a coherent book out of a massive jumble of myths, rumors, and anecdotes. Frank's semi-infinite patience and his love of correct writing result in a very satisfying product. Credit is also due to Alexandra Simon, copy editor at Facts On File, for her superlative job of finessing and polishing the manuscript. The support and the editing skills of my wife, Carolyn, were also essential. She held up the financial life of the household while I wrote, and she tried to make sure that everything was spelled correctly, all sentences were punctuated, and the narrative made sense to a nonscientist.

Introduction

The discovery and application of nuclear power were among the most profound scientific accomplishments of the 20th century, beginning with tentative explorations of the structure of matter, expanding into a rapid succession of unexpected discoveries, and finally settling into a seamless transition from theoretical science to applied engineering. In that century everything changed, as follows:

❋ Science changed from an academic pursuit to an industry.
❋ The scale of mathematical modeling changed from predicting the action of a bouncing ball to predicting the actions of trillions of simultaneously bouncing neutrons.
❋ The use of *uranium* changed from an occasional orange or green dye in ceramics to a major power fuel.
❋ The concerns of public safety changed from boiler explosions on steamboats to nuclear *reactor* explosions on continents.
❋ The concept of warfare changed from endangering soldiers on battlefields to endangering populations in cities.

The History of Nuclear Power describes the sequence of these changes, as science and technology rapidly matured more than 100 years and as the scale of civilization and its energy needs expanded. Sidebars supplement the historical narrative, providing interesting notes on many of the pioneering scientists involved in the development of nuclear science and technology, as well as notes on spin-off ideas and branching technologies. The narrative follows the pace of nuclear development, surging ahead faster than civilization could keep up with it, stumbling occasionally, finally pausing to assess and contemplate all that had been accomplished, and never looking back.

Nuclear power in the United States was in a quiescent state for three decades, neither developing forward nor shutting down, and delivering about 20 percent of the nation's electrical power. Other technologies, such as electronics, computers, and communications systems, rushed ahead in this period, improving and innovating. The art of generating power by nuclear means stood stagnant, trying only to make electricity while remaining out of the public eye. The situation is now changing in complex ways. There is a heightened awareness of global climate shifts, the chemical composition of air, and the finite nature of burnable fuels. These new concerns would seem to favor a renewed push for nuclear power production, among other nonpolluting methods, but there are multiple layers of public anxiety. We are worried about future weather patterns and a lack of gasoline, but we are also worried about long-lived radioactive contamination and the safety of nuclear reactor operations. As these issues are pondered, a heightened level of understanding of nuclear science and its applications will be important enough to affect career paths and college majors.

The History of Nuclear Power provides a fundamental introduction to this complicated subject. It follows a straight line down the middle of the larger subject of nuclear technology, concentrating on the development of light-water *fission* reactors as the dominant power source design, skirting other interesting technologies, such as hydrogen *fusion* reactors or space propulsion reactors. These and other important topics are covered in further volumes in the Nuclear Power multivolume set.

I have been taught the history of nuclear power by its participants. My graduate school professors in nuclear engineering worked on the *atomic bomb* project during World War II, the nuclear-powered strategic bomber, the nuclear rocket engines, and the space-borne power reactors. I entered the workplace just as these projects were disappearing over the horizon, but I found a new set of frontiers and participated in the second phase of the history of nuclear power. I bring my experience and the knowledge passed from my elders to this work, and I hope that you will find it fascinating.

Nuclear technology must be approached with an enhanced sense of industrial safety, unprecedented in the history of mechanical systems, and the issue of nuclear hazards will be present in any discussion or debate on nuclear subjects. *The History of Nuclear Power* demonstrates the speed with which it was necessary to adjust industrial mind-sets to this new level of safety consciousness, and specifically dangerous aspects of the technology will be treated in detail in further volumes of the series. *The History of Nuclear Power* also reveals the sudden shift in the center of gravity of the body of nuclear science to the United States immediately before World War II, as the world's top scientists fled their homelands and universities in Europe to escape troubling political developments. This fortuitous concentration of genius in the United States, which was seen as an island of freedom and safety in an unsafe world, led to an unusually rapid development of nuclear technology. Unique aspects of this development were the military takeover of all nuclear science during World War II and the smooth transition from fanciful theories to working industrial systems and weapons of immense power. After the war, through creative engineering, important legislation, and political arm-twisting, this new weapons technology was transformed into a peaceful, civilian-controlled energy source. Such is the first century of nuclear power development. The second century may require a similar quantity of groundbreaking science, advanced engineering, statesmanship, global diplomacy, and an ability to plan for the future.

The History of Nuclear Power has been written as a stirring account of the genius, the hard work, and the pure luck needed to unlock the atomic *nucleus* and turn matter into energy for the student or the teacher who is interested in seeing the future through a study of the past. Technical details of the nuclear process are made understandable through clear explanations of terms and expressions used almost exclusively in nuclear science. Much of nuclear technology still uses the traditional, American system of units, with some archaic terms remaining in use. The cross-sectional area of a nucleus, for example, is still universally and officially expressed in barns, and not in square centimeters, due to a purely historical fluke. An American scientist, upon first measuring the cross section of a uranium nucleus, exclaimed, "That's as big as a barn!" Where appropriate, units are expressed in the international system, or SI, along with the American system. A glossary, chronology, and a list of current sources for further reading and research are included in the back matter.

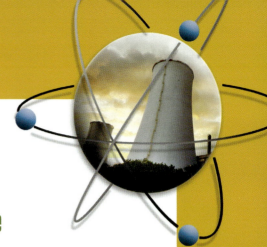

1 Centuries of Atomic Structure Theories

One of humankind's first scientific discoveries was fire. At some distant unrecorded date it was found that dead organic matter, such as tree limbs, could be made to burn, and human beings grew to enjoy cooked food, lighted shelters, and a warmth that allowed comfortable living in cold climates.

Breaking the weak forces that hold the *atoms* of a stick of wood together makes heat. Breaking the powerful forces that hold the nuclei of a stick of uranium together makes millions of times more heat than simple burning. This principle of nuclear power is now well understood, scientifically accepted, and widely practiced, but it was a long effort to get to this point of knowledge. Before the power of the nucleus could be explored, or even contemplated, it was necessary to realize that matter is divided into atoms.

This chapter first will show the gradual realization of atomic structure, starting as a hypothetical philosophy in ancient times and eventually refining into more rigorous, practical theories in the 19th century, as the concept of matter divided into indestructible chemical elements became clear and the practice of formal science was established. The discussion then reveals that when it was confirmed that atoms can neither be created nor destroyed, it was found, by accident and experimentation, that pieces of an atom can be torn off, and various forms of radiation result from this action. Light and the newly discovered radio waves and *X-rays* were

1

found to be different manifestations of the same phenomenon, which is an electromagnetic radiation predicted to exist by a set of finely crafted mathematical equations. The chapter goes on to study the alarming discoveries near the end of the 19th century, when an additional source of a more powerful radiation was found, apparently coming from deep inside the atom and requiring no external stimulus.

EARLIEST CONCEPTS OF ATOMIC STRUCTURE

There has always been a need to analyze things and substances down to component parts in order to explain material characteristics in terms of combinations of some simpler, basic pieces. Near the beginning of civilization, as writing, fixed agriculture, and manufacturing became human activities, a common theory of element analysis seemed to appear in several places. This practical, working theory was that everything is composed of various combinations of four elements: earth, air, fire, and water. Although this concept now seems quaint, in ancient times it made a certain logical sense. Steam, for example, was obviously composed of air, containing a measure of water, giving it wetness, plus fire, giving it heat. Bricks were made of earth, with the water removed, wine was water with a bit of earth and fire mixed in, and something as complex as wood was mainly earth, with some water, air, and fire locked in, to be extracted when the wood was burned. Burn the wood, and the fire would escape, the water and air would evaporate away, and one is left with only a pile of black earth or ashes.

With this rough but practical working theory, technology and science managed to progress very slowly for thousands of years. There were some other theories, often showing brilliant insight in a world lacking a base of scientific knowledge. The first written mention of a true atomic analysis of matter dates to around 550 B.C.E. in India, where elaborate theories were developed by the Nyaya and Vaisheshika schools, describing how elementary particles combine, first in pairs, then in trios of pairs, to produce more complex substances. The first references to an atomic structure in the West appeared 100 years later. A teacher named Leucippus (ca. fifth century B.C.E.) in Greece thought of a scheme in which all matter was composed of smaller pieces, with the smallest pieces being incapable of being broken into smaller pieces. His views were recorded and systematized by a student, Democritus (ca. 460 B.C.E.–370 B.C.E.), around 430 B.C.E., and in this work the word *atomos* was first used, meaning "uncuttable." The Greek word was later shortened to *atom*. These were fine theories and were pointed in the right direction, but they were of no practical use and were considered philosophy. These theoretical atoms were too small to be seen, and there was no experimental confirmation that any of these ideas had a basis in reality.

Through the turn of the millennium, in 1000 C.E., the practice of industrial chemistry, in which useful compounds such as soap were formed by mixing two or more substances, increased in importance, and the lack of utility in the ancient earth-air-fire-water model of matter began to become evident. Gold, for example, was apparently earth because it was a solid, but the difference between gold and copper was difficult to define or explain. There was no systematic method of analysis available and no way to quantify the subtle differences among metals, liquids, or gases. Any analysis was simply an opinion, and the art of chemistry had to turn to unproductive, mystic explanations for compounds and alloys.

In 1661, Robert Boyle (1627–91), a well-educated, Irish gentleman of independent means, having attended Eton College in England, published a book with the verbose title *THE SCEPTICAL CHYMIST: OR CHYMICO-PHYSICAL Doubts & Paradoxes, Touching the SPAGYRIST'S PRINCIPLES Commonly call'd HYPOSTATICAL, As they are wont to be Propos'd and Defended by the Generality of ALCHYMISTS.* This book broke new ground in that Boyle finally called the four-element system to task and harshly criticized the more advanced work of the Spagyrists, who contended that solid matter was composed of various combinations of salt, sulfur, and mercury. He went further to advance the theory, once again, that matter is composed of elements, which are undecomposable atoms, and he described the process of chemical analysis and

Robert Boyle, a founder of modern chemistry, ca. 1689 *(Granger Collection, New York)*

the fundamental differences between compounds and mixtures of compounds. Boyle's other work studying the properties of gases is considered an important beginning to formal science, but his chemistry book was a monumental start in the understanding of atomic structures.

After Boyle, formal science gathered speed, and the ancient concepts of matter were buried. In 1789, Antoine-Laurent Lavoisier (1743–94), a French nobleman, chemist, and economist, first used the term *elements* to describe oxygen and hydrogen, claiming that these two gases could not be broken chemically into more elementary components. He compiled the first table of elements and went on to introduce the metric system of measurements. He also advanced physics by formulating the law of conservation of mass, by which matter can neither be created from nothing nor destroyed but only be changed in form.

In 1808, John Dalton (1766–1844), an English chemist, meteorologist, and physicist, published the first volume of his *New System of Chemical*

A reconstruction of the laboratory of Antoine-Laurent Lavoisier, a Frenchman who compiled the first list of elements *(Foto Deutsches Museum)*

Philosophy, in which he stated the following five main points of his atomic theory:

- ❊ Elements are composed of indivisible particles called *atoms.*
- ❊ All atoms of a given element are identical.
- ❊ The atoms of a given element are different from those of any other element.
- ❊ Atoms of one element may combine with the atoms of other elements to form compounds.
- ❊ Atoms may not be broken into smaller particles, destroyed, or created from combinations of smaller particles by chemical action.

Although these simple rules may now seem obvious, Dalton's work solidified Boyle's findings and set the course for chemistry and physics for the next 200 years.

By the late 19th century, the existence and the importance of the atom were firmly established. The next increment of knowledge would be large and unexpected, when it was discovered that the undecomposable, indivisible atoms were falling apart.

FLUORESCENCE AND THE DISCOVERY OF RADIOACTIVITY

The next steps in the development of atomic theory were the discovery of mysterious *electromagnetic waves* that could not be seen with the naked eye and an eventual realization that all these waves, regardless of the means used to produce them, were of similar character and were the result of activity within the atom.

The investigation of electromagnetic waves started appropriately, with theoretical predictions of their existence. The first suggestion of electromagnetic radiation was from an English chemist and physicist named Michael Faraday (1791–1867), who in 1831 started experimenting with electromagnets. Faraday found that a changing magnetic field produces an electric field, and that he could induce electricity in a nearby magnetic coil using a changing magnetic field. Faraday went so far as to propose that electromagnetic forces extended into the empty space surrounding one of his electromagnets, but the idea was roundly rejected by his fellow scientists.

EVIDENCE OF PREHISTORIC NUCLEAR ACTIVITY

Two definite milestones in the history of nuclear power were the manufacture of plutonium, the first man-made element, in 1941, and the first sustained nuclear reaction in 1942. Both milestones were modified in 1972 when it was discovered that there had been an operating nuclear power reactor 1.5 billion years ago, and that it had produced 3,300 pounds (1,500 kg) of plutonium. There were actually 16 reactors, which ran for a few hundred thousand years, breaking all run-time records and producing energy at an average rate of 100 kilowatts, in the Oklo uranium mine, in Gabon, Africa.

This remarkable discovery was made at the Pierrelatte Uranium Enrichment Facility in France. Output from uranium mines was routinely analyzed by a mass spectrometer, to insure that every atom of uranium fuel was accounted for and none were being diverted for weapons production. In May 1972, samples from the Oklo mine showed a curious discrepancy. Particular attention was given to the *fissionable isotope U-235*. The normal concentration of U-235 in raw uranium is 0.7202 percent. The Oklo samples showed only 0.7171 percent, and the difference was significant. The French Commissariat à l'énergie atomique launched an immediate investigation, finding concentrations of U-235 as low as 0.440 percent in the Oklo uranium.

Clues from a detailed analysis of the mineral content of the mine led to a startling conclusion. On September 25, 1972, the Commissariat announced their finding: Self-sustaining nuclear chain reactions had occurred at the Oklo uranium mine about 1.5 billion years ago, producing 12,000 pounds (5,400 kg) of fission waste products and depleting the fissionable uranium in the ore.

The natural reactors formed in uranium-rich mineral deposits when groundwater inundated the ore. The water acted as a neutron moderator, bringing the concentrated uranium deposits to criticality, raising the temperature of the ore to a few hundred degrees Celsius, and boiling the water. As the water boiled away, a natural reactor would shut down, resulting in a pulsed action, over an interval of about 2.5 hours, as water once again collected in the ore, repeating the process for 100,000 years. At the time, more than a billion years ago, the U-235 concentration in the ore was about 3 percent, which is comparable to the fuel used in some power reactors today. Since the U-235 component decays away faster than the remainder of the uranium ore, the concentration of U-235 in natural uranium has dropped to about 0.7 percent since the natural reactors last powered up.

By 1864, the concept of electromagnetism in space was reconsidered by a Scottish mathematician and theoretical physicist named James Clerk Maxwell (1831–79). Faraday's knowledge of algebra had been weak, and he could not formulate a mathematical argument for his idea, but Maxwell was a genius at calculus and had earned the Second Wrangler of Mathematics degree at Trinity College, Cambridge. Maxwell was interested in everything scientific. He wrote an original essay in college, "On the Stability of Saturn's Rings," in which he concluded that the rings were not completely solid, nor liquid, but were composed of "brickbats." He did

Michael Faraday, an English chemist and physicist, in his basement laboratory in 1852 *(The Royal Institution, London, U.K./The Bridgeman Art Library)*

some important work on color and color blindness and took the world's first color photograph in 1861, of a Scottish tartan. He studied Faraday's work on magnetic lines of force, and with that as an inspiration, he formulated a set of 20 differential equations, in 20 variables describing the magnetic and electrical fields in both static and dynamic conditions.

The equations were complicated and difficult to fathom, but in these equations was a perfect, mathematical prediction that there exist waves of oscillating electric and magnetic fields that travel through empty space at a predictable speed. The speed predicted happened to be the speed of light, and Maxwell jumped to the conclusion that light is an electromagnetic wave, vibrating in a frequency band that we can detect with our eyes. Maxwell would be proven correct.

The implications of Maxwell's equations remained an elegant but unapplied theory until Heinrich Rudolf Hertz (1857–94), a German mathematician and physicist, made an accidental discovery in 1887. Hertz earned his Ph.D. in 1880 at the University of Berlin and became a full professor at the University of Karlsruhe in 1885. He had dabbled in the investigation of many subjects, including meteorology and elasticity, but in 1887 he was working with a newly invented piece of high-tech equipment. It was a high-voltage coil, producing sparks a half-inch long, with a buzzer built into the end of the coil to sustain the spark. Hertz was fascinated by the effect of light on the spark. He noticed that the spark seemed to dim when *ultraviolet light* hit it. The light was apparently knocking electrical charge off the spark gap, and this was an exciting finding.

Of even greater importance than this photoelectric effect was an unexpected by-product of the high-voltage spark. As Hertz turned off the lights to get a better look at his spark under ultraviolet, he noticed something out of the corner of his eye. There was another spark occurring in the room, in the gap between the ends of a loop of wire that was not connected to the apparatus. To his amazement, the spark produced by his high-voltage coil was somehow perceived and replicated by another spark gap, sitting on another table in the room. This concept of action at a distance seemed profoundly strange. There were no electrical wires connecting the two pieces of equipment, and yet if he threw the switch on his spark coil, a spark would occur on a loop of wire on the other side of the room. He was affecting the loop of wire, the antenna, by generating Maxwell's electromagnetic wave. Hertz had discovered radio, and he had confirmed Maxwell's vision of radiating waves.

Wilhelm Roentgen (1845–1923), a German physicist, was also fascinated by the high-voltage coil and its novel effects. Roentgen had graduated from the University of Zurich in 1869 with a Ph.D. and was named the physics chair at the University of Würzburg in 1888. He was studying the effects of a high-voltage coil connected to an evacuated glass tube. Study of the newly discovered *cathode rays* was popular in Europe in the 1890s, and it seemed as if everyone on the leading edge of science was experimenting with some form of vacuum tube. The *cathode* ray would soon be identified as a stream of *electrons,* or small components of atoms stripped off by high voltage, but in 1895 the ray was only known to travel from one end of the tube to another, from the negative to the positive high-voltage electrodes, causing the glass to fluoresce. Roentgen wanted to find out if he could cause the rays to leave the tube and enter the air surrounding it. In the late afternoon of November 8, 1895, he tried a special tube, built by a colleague, having a thin, aluminum window on the end. The cathode rays might penetrate the aluminum, and he would use a piece of cardboard painted with barium platinocyanide as a detector.

Roentgen's X-ray Tube

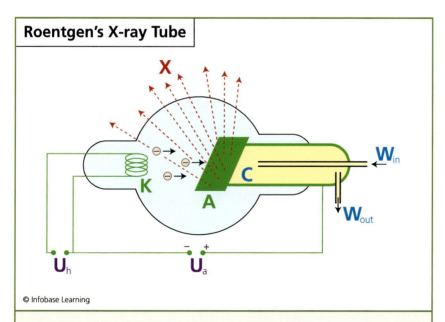

© Infobase Learning

X-ray apparatus is encased in a glass vacuum tube. **K** is the cathode, a metal filament heated by an electric current, U_h. The anode is **A**, cooled by water in the sealed vessel **C**. W_{in} is cooling water in, and W_{out} is heated water out. U_a is a high-voltage direct current applied across the cathode and the anode. Electrons leaving the hot cathode at high speed crash into the anode, where the rapid deceleration causes X-rays, **X**, to leave the tube.

Being careful, Roentgen devised a cardboard shield to fit over the tube so that no fluorescent light would escape and spoil his measurement, but as he dimmed the lights in the laboratory to test his shield with the tube running at full power, he noticed something out of the corner of his eye. Just as Hertz had noticed his sparks, Roentgen noticed that his piece of cardboard, on a lab bench more than a meter away, was shimmering with yellow-green light. He had hoped to get cathode rays out of the tube, but he knew that they could not have enough energy to bore through the air and hit the barium screen that far away. He had discovered a new type of ray. When the cathode rays hit the aluminum window at the positive electrode end of the tube, they were stopped, and the sudden deceleration produced high-energy rays, invisible and streaming out the end of the tube, just as Maxwell's equations had predicted. Experiments over the next few days proved that these new rays were more powerful than light and could penetrate solid objects. Needing a quick, temporary name for his discovery, Roentgen called them X-rays.

By 1896, atomic science was progressing rapidly, with physics journals having trouble keeping up with the rate of discovery. Antoine-Henri Becquerel (1852–1908), a French physicist, was caught up in the excitement and was investigating the work of Wilhelm Roentgen. Although he had studied physics at the École Polytechnique, there were practical considerations for getting a paying job, so he also studied engineering at the École des Pont et Chaussées and became chief engineer in the Department of Bridges and Highways.

Practical work did not keep him from his fascination with Roentgen's work, which was very successful, with immediate applications in medicine, but not completely understood. The composition of cathode rays was unknown. It was known only that something would stream from the negative electrode, or cathode, at one end of a glass tube, with the air removed, to the positive electrode at the other end of a glass tube, when 30,000 volts were applied to the electrodes. When the cathode rays hit the glass at the positive end, they caused the glass to glow, but, aside from that, the cathode rays were invisible in a hard vacuum. Roentgen still did not realize that his X-rays were produced by electrons hitting his big, aluminum, positive electrode, because the electron had yet to be discovered. Becquerel went to the weekly meeting at the muséum national d'Histoire naturelle in Paris on January 20, 1896, to hear a report on Roentgen's work in Germany. Roentgen was convinced that his powerful X-rays, which would penetrate light-shielding and fog photographic plates, were produced by the induced *fluorescence* in the end of the tube.

It occurred to Becquerel that if the weak fluorescent glow at the end of a cathode-ray tube produced X-rays, then he could produce a greater flux of X-rays by using a material that would give a bright, robust fluorescence under ultraviolet light. He immediately bought all the fluorescent materials he could find and began experimenting, using the ultraviolet component of sunlight to excite fluorescence and using sealed photographic plates to record his X-ray production. Although his experiments were carefully assembled, he was getting no results. In 10 days of experimenting, he could not fog any film with fluorescence-induced X-rays. On January 30, he read an article on X-rays, and it encouraged him to keep trying.

Becquerel bought some *uranium* salt, uranyl potassium sulfate, the most strongly fluorescent substance available, sprinkled some atop a sealed photographic plate and exposed it to sunlight for several hours. The experiment was immediately successful, or so he thought. When he developed the plate, he could see the black silhouette of the sprinkled uranium salt on the negative. Obviously, he had found the right fluorescent material to make X-rays using sunlight. The commercial possibilities of the discovery were wonderful. He could manufacture a simple medical X-ray machine that would require no electricity and no fragile glass tubes and could be used in remote locations.

Just to make sure of the results, on February 26, Becquerel prepared another photographic plate, wrapped in thick, black paper, with a small amount of uranium salt on top. Unfortunately, the weather in Paris had turned cloudy. With no sunlight, he slipped his experiment into a dark drawer in his desk. The next day was cloudy as well. On March 1, for some odd, serendipitous reason, Becquerel decided to go ahead and develop the plate, without any ultraviolet light having excited the fluorescent uranium.

To Becquerel's amazement, the plate was clouded, as if the light-shield had been defective, but the shape of the dark cloud was a perfect replica of the irregular scattering of uranium salt. Furthermore, the clouding on a plate abandoned in a dark drawer for three days was much darker than he had achieved in sunlight for a few hours. He started putting the evidence together, and he realized that the sunlight and the fluorescence had nothing to do with the effect. It was something in the uranium that was clouding the plates. Henri Becquerel had discovered some kind of force that could cloud a photographic negative, through the light-tight cover, requiring no high-voltage tube to produce it. It was something that could not be felt, seen, heard, tasted, or smelled. He gave it a name: Becquerel rays.

In a few years, Becquerel's important discovery would be given a new designation by Marie Curie (1867–1934), *radioactivity*.

PROOF THAT ATOMS CAN BE BROKEN

Sir Joseph John "J. J." Thomson (1856–1940) was born in Manchester, England. Showing early interest in technical matters, he studied engineering at the University of Manchester in 1870 and then moved to Trinity College, Cambridge, in 1876 to study mathematics. In 1880, he earned a B.A. degree (Second Wrangler) and an M.A. in 1883. In 1884, he became Cavendish Professor of Physics, in 1890, he married the daughter of the Regius Professor of Physics at Cambridge, and in 1897, he analyzed the atom into component parts, sending atomic science bounding in new directions.

Thomson was interested, as were many of his fellow physicists, in the mystery of the cathode rays. He built more sophisticated, more complicated glass tubes, in which he electrically accelerated the ray from the tube's negative electrode through holes drilled in positive electrodes, sending the beam gliding through the deep vacuum beyond the electrodes and to the far end of the tube, where it would hit a fluorescent screen and cause a small spot to glow. He found that he could deflect the thin cathode ray streaming through the hole in the positive electrode using a magnet at the side of the tube.

To investigate the nature of the cathode rays, Thomson devised three, sequential experiments. The cathode rays obviously involved a negative charge, as they originated at the negative electrode and vanished into the positive electrode, and for his first experiment Thomson wanted to know whether the negative charge could be separated from the rays. He built a special variant of his tube, blowing a thin, wide beam of cathode rays through a slit in the positive electrode. This beam would traverse the tube, unencumbered by air molecules, and hit a third electrode at the end of the tube. He connected an electrometer to the electrode to measure the charge from the cathode rays and confirmed that there was an electrical current flowing between the negative electrode origin of the rays and his target electrode. The target electrode had a slit cut in it, off the straight axis of the beam. With the tube operating at full power, Thomson adjusted a horseshoe magnet across the length of the ray's flight path,

Sir J. J. Thomson, British discoverer of the electron, in 1904, studying the behavior of cathode rays in the Cavendish Laboratory, Cambridge, England *(Granger Collection, New York)*

throwing the beam into a downward turn. He aimed it for the hole in the third electrode. The rays missed his electrode and hit the wall of the tube, causing fluorescence. At that point, the electrical current stopped registering on his electrometer. Thomson concluded that the electrical charge and the rays were one and the same, and that one could not be separated from the other.

Thomson suspected that he knew the nature of the mysterious rays, but he set up experiment number two for a stronger case. If the rays were purely electrical charge in motion, then he should be able to bend the rays with a stationary electrical charge. He set up another tube, this time with a thin beam established at one end of the tube and shot through a couple of parallel metal plates, against a fluorescent screen at the far end of the tube. This experiment had been tried several times by others with no results, but Thomson thought he knew why. The ray must have been crashing into gas particles left in the tubes, because of imperfect vacuums. Thomson made

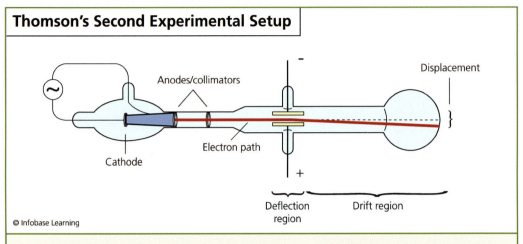

Thomson's Second Experimental Setup

J. J. Thomson's vacuum apparatus projects a thin stream of cathode rays down the length of the tube. By demonstrating that the stream could be bent using an electric field, Thomson proved that the stream consisted of negatively charged particles.

certain that his tube was pumped all the way down. He put the 30,000 volts on the negative and positive electrodes, turned down the lights, and observed his bright spot on the end of the tube, as he charged the parallel plates from a battery. Just as he had thought, the beam deflected away from a negative electrical charge and toward a positively charged plate. He could not see the beam itself, but he could watch as the spot of light changed position on the end of the tube. Knowing the angle of deflection of the beam and the voltage required to do it, he was able to calculate the ratio of charge to mass of the particle he suspected made up the beam.

There was one more experiment needed. Thomson repeated the beam-deflection measurement using a magnetic field instead of an electrical field to bend the beam, and again he calculated the ratio. He was then prepared to make a bold, sweeping conclusion: The cathode rays were composed of tiny negatively charged particles he called "corpuscles," which were stripped-off atoms in the negative electrode and thrown down the length of the tube. He went further, to propose that, because matter was naturally without electrical charge, the rest of the atom, with the electrons stripped off, had to be positively charged, so that it would cancel the negative charge of his corpuscles. He imagined that the tiny lightweight electrons were stuck in a relatively large, soft ball of positive charge. It was called the "plum pudding" model of the atom, and it would do nicely for the time being.

J. J. Thomson's corpuscles would later be named electrons, and he would be awarded the Nobel Prize in physics in 1906 for this important discovery.

MARIE AND PIERRE CURIE FIND RADIUM IN URANIUM ORE

Maria "Manya" Skłodowska (1867–1934) was born in Warsaw, then a part of Poland under the occupation of the Russian Empire. As a child she was encouraged to seek a higher education by her mother, a math teacher, and her father, a physics teacher, and eventually she was able to attend the Floating University, an illegal night school in Warsaw. Working as a tutor and as a governess for children of wealthy families while studying mathematics and chemistry, Manya was eventually able to gain acceptance to the prestigious Sorbonne. In 1891, she moved to Paris and changed her name to Marie, to fit into the French culture, as she applied herself diligently to her studies in math and physics.

By 1894, Marie had performed pioneering research on magnetism and steel, and she was the laboratory chief at the Municipal School of Industrial Physics and Chemistry in Paris, where she shared laboratory space with a like-minded scientist named Pierre Curie (1859–1906). In July 1895, the two scientists were married, and Marie Skłodowska became Marie Curie. The research work of Marie and Pierre Curie was performed in a barely adequate structure in Paris, fondly referred to as "the miserable old shed," with minimum funding, and yet they were able to steer the course of atomic science and be awarded three Nobel Prizes between them. Marie was the first person to win Nobel recognition in two different sciences, physics and chemistry. The 1903 Nobel Prize in physics was shared by Marie, Pierre, and Marie's doctoral thesis adviser, Henri Becquerel.

In 1896, Becquerel's newly discovered rays were considered interesting by the scientific community, but much more attention was focused on Wilhelm Roentgen's X-rays. Marie found the neglected rays from uranium interesting, and she used a new technique to detect and quantify them. A precision electrometer had been invented 15 years earlier by her husband, Pierre, and his brother, Jacques. She used it to measure the ionization effect in air caused by the passage of Becquerel rays. Using this novel equipment setup, she was able to confirm Becquerel's observations that the radiation from uranium is constant, regardless of whether the uranium was solid or pulverized, pure or in a compound, wet or dry, or

exposed to light or heat. Minerals having the highest concentration of uranium seemed to emit the most radiation. She then went farther than Becquerel, suggesting a hypothesis that the rays were a result of some property of the very structure of the uranium atoms.

In 1898, Marie found another element that emitted Becquerel rays. It was thorium, and she was becoming convinced that it was an atomic property and not some external cosmic-ray influence. By this time, Pierre was so intrigued by Marie's findings that he dropped his own investigations into crystals and joined her in studying pitchblende and chalcolite, which were uranium ores. She had found something interesting: Pitchblende, from which uranium was extracted, was more radioactive than was pure uranium. There was apparently something mixed in with the uranium.

By chemically processing tons of pitchblende, the Curies were able to identify two new radioactive elements existing in the same mineral with uranium. The first element discovered Marie named *polonium,* in honor of her native Poland. The second she named *radium,* for its aggressive radioactivity. With tremendous difficulty, Marie and Pierre managed to refine the radium down to a pure metal, in sub-gram quantities. It was an interesting material. It would glow blue in the darkened laboratory, but the power and the danger in that blue glow were only suspected.

(opposite) Marie (1867–1934) and Pierre (1859–1906) Curie, codiscoverers of radium, as they appeared in *Vanity Fair* magazine in 1904. Marie is in her blue wedding dress, which she wore for many years as a lab coat. She stands behind Pierre, who was until his death always listed as the principal investigator, holding in his left hand a dangerously radioactive sample, as usual. *(National Library of Medicine)*

2 Discovery of the Atomic Nucleus

By 1899, scientific studies had established that matter is divided into characteristic atoms and that electrically charged components of these atoms can be ripped off and sent flying through a vacuum. These tiny, invisible components, later to be named electrons, behave as predicted by the mathematical models formulated by Maxwell and Faraday. The flight of an electron through the vacuum deflects in a predictable trajectory by an imposed magnetic or electrical field. Furthermore, Maxwell and his groundbreaking set of equations had predicted that a changing magnetic field causes an electrical field and that a changing electrical field causes a magnetic field. Strike a high-voltage spark across two electrodes and the burst of an electrical field will cause a magnetic field, which causes another electrical field, which causes another magnetic field, and so on, as a wave of electrical and magnetic fields radiates through space at the speed of light, alternating between electricity and magnetism at a frequency that is proportional to its energy of creation.

The gradual discovery and confirmation of this electromagnetic wave phenomenon would prove monumental, as it became evident that light itself was a manifestation of this wave effect. A wave of lesser energy would be exploited for radio communications, and waves of greater energy would be used as X-rays for medical diagnosis. These interesting scientific discoveries would spin off into successful commercial products in the new century, but the scientists would continue to push open the door of discovery, ever curious concerning the nature of matter and finding that solving a puzzle of the natural world simply uncovered more puzzles. Scientists in Germany and France found that there were other ways to derive radiation without direct application of the Maxwell equations. Some heavy elements, such as uranium and the newly discovered polonium and radium, would dismantle themselves on the atomic level, emitting even more powerful forms of radiation.

ERNEST RUTHERFORD STARTS NAMING RAYS AND PARTICLES

In 1898, Ernest Rutherford (1871–1937), a scientifically talented young man from New Zealand, studied the radiations emitted from the elements uranium and thorium. Working at the Cavendish Laboratory of the University of Cambridge, he found two distinct types of radiation, and he named them. The first seemed to have little range. It was easily stopped by air or by thin barriers of almost anything solid, and he named it alpha radiation. The second type had greater range in air and was better at penetrating shields. Rutherford named it beta radiation. A few months later, Paul Villard (1860–1934), working in the chemistry department at the École Normale in Paris, identified a third, even more penetrating radiation type emitting from uranium. In keeping with Rutherford's newly established naming convention, he called it gamma radiation.

In 1898, when he was 27 years old, Rutherford moved to Canada to become professor of physics at McGill University in Montreal. Here he had a new, well-equipped physics laboratory, generous funding, and a learned colleague in chemistry named Frederick Soddy (1877–1956). Almost immediately upon arrival, Rutherford presented Soddy with a puzzle: There was some sort of gas emanating from radioactive thorium. What might it be? A chemical analysis was in order.

Soddy analyzed the sample and found that the gas had no chemical characteristics whatsoever. There was only one conclusion possible, that the gas was an inert chemical such as argon. Odd as it seemed, the element thorium was apparently transmuting itself into argon gas, slowly but steadily. This discovery of the spontaneous disintegration of radioactive elements was a major discovery, and Rutherford and Soddy immediately investigated the known radioactive elements to discover what was happening. By literally counting the number of radioactive particles emitted from a sample during a given time, they found that each radioactive

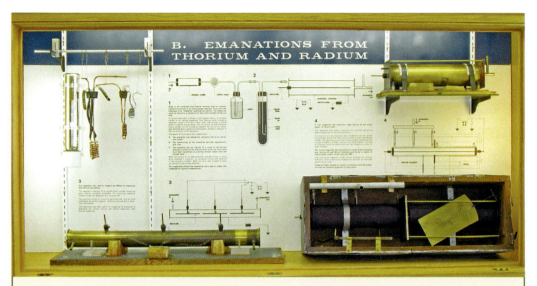

Ernest, Lord Rutherford's apparatus from his "Emanations from Thorium and Radium" experiment, on display at McGill University, Montreal, Canada *(Rutherford Museum, McGill University)*

substance was decaying exponentially, at a characteristic rate, or that the source of the radiation would drop by half in a predictable passage of time. Soddy named the characteristic time *half-life,* meaning the time required for the radioactivity to decrease by half. For a given sample of a radioactive substance, the radiation level would drop by half in one half-life. In another half-life, what was left of the radioactivity would drop by another half, and so on, forever. The radioactivity would never technically disappear, but it would drop by halves in a predictable time period.

Rutherford suspected that *beta rays* were, in fact, a naturally occurring form of cathode rays being generated by many of his colleagues using electrically stimulated vacuum tubes. He was correct, and he demonstrated it using magnetic and electrostatic fields to bend beta rays in a vacuum tube. Instead of using high-voltage electricity between electrodes in the tube, he simply put a sample of uranium at one end. He also suspected that alpha rays were actually helium atoms stripped of their electrons, and he was able to test that theory in a most elegant way at the University of Manchester in England in 1908.

Rutherford's proof of the nature of alpha rays was stunning for its simplicity and almost artistic style. He had a glassblower make him a tube with walls thin enough for alpha rays to penetrate. The tube was evacuated, filled with radon gas (a known alpha-ray emitter), and sealed off at the end. This tube was then put inside another, larger tube with thick walls, which was pumped down and flame-sealed at the end. Rutherford used a light spectrometer to detect anything in the vacuum between the tubes. There was nothing there. He waited a few days and tried again. The space between the tubes had become filled with helium. Therefore, the alpha rays were actually positively charged helium ions, broken free of the much heavier radon and thrown through the thin glass of the inner tube. The name of the radiation was adjusted, from alpha rays to *alpha particles,* and Rutherford noted that this demonstration explained why helium is found trapped in the crystalline spaces in thorium and uranium ores. He announced the triumphant finding to the audience in Stockholm as he accepted his Nobel Prize in chemistry. Soddy had been almost right about his analysis of the mysterious decay product of thorium. It was not argon. It was another inert gas, helium.

THE ENERGY RELEASED BY RADIOACTIVE DECAY

In 1903, Rutherford collaborated with Frederick Soddy to write an important paper, "Radioactive Change." In this work they offered the first experimentally verified calculations of the energy released from an atom due to radioactive decay. The power involved in the transmutation of radioactive elements was astounding. They had found that the energy released by the decay of one gram of radium could not be less than 100,000,000 gram calories. It was probably closer to 10,000,000,000 or 10 billion gram calories.

In 1903, at the University of Kiel in Germany, Philipp Lenard (1862–1947) reached an interesting conclusion regarding atomic structure. Rutherford was in accordance with J. J. Thomson's opinion that the atom was one solid mass, like a plum pudding, with electrons adhering to the outside, remarking that, "I was brought up to look at the atom as a nice hard fellow, red or gray in color, according to taste." Thomson was, after all, his thesis adviser for his doctorate, awarded in 1900. A solid object, such as a block of metal, was obviously hard, massive, opaque, continuous, and homogeneous.

Lenard had been working on *cathode ray tubes,* hoping to accomplish what Roentgen had tried, bringing cathode rays out the end of the glass vacuum tube and into the laboratory. He had devised a metal window thick enough to withstand the air pressure outside the tube but thin enough for cathode rays to penetrate and flow into the atmosphere. It worked. He was able to detect cathode rays outside the tube using a fluorescent screen, but he noticed that the rays were scattered somewhat when they blew through the metal window. This seemed to contradict the theory that they were

ERNEST RUTHERFORD:
THE MAN WHO SORTED OUT THE ATOMIC STRUCTURE

Ernest Rutherford, first baron Rutherford of Nelson, OM PC PRS, or, simply, Lord Rutherford, invented the discipline of *nuclear physics* by discovering the atomic nucleus and is considered a primary pioneer in the field of nuclear research.

Rutherford was born near the town of Nelson, New Zealand. His parents had moved there from Perth, Scotland, to raise flax and children when New Zealand was still a rough frontier outpost of the British Empire. Young Ernest won academic scholarships, first to Nelson College and then to the University of New Zealand. After earning his BA, MA, and BSc in 1893, he performed two years of research at the university, looking into Hertz's 1887 discovery of electromagnetic radiation from a spark gap.

Impressed with his work on Hertzian oscillators, Cambridge University in England offered him a scholarship. His mother gave him the triumphant news from a received telegram, shouting it to him as he dug up potatoes in the family garden. Rutherford reportedly tossed away his spade, exclaiming, "That's the last potato I'll dig!" He was correct. His genuine genius, his ability to be continuously astonished, and his country-boy ability to improvise would converge in one of the finest talents ever in experimental physics, right in the middle of an exciting time in science when the basis of matter and energy required analysis. He moved to Cambridge in 1895 to work with the director J. J. Thomson, and they soon turned to investigations of the radioactivity discovered by Becquerel, the Curies, and Roentgen. Over the next decades of his work, Rutherford would systematically dissect the atomic structure, discover the nucleus, and win the Nobel Prize in chemistry in 1905.

Rutherford's most famous saying was made in 1933, when he was quoted in a newspaper article commenting offhand that any attempt to derive usable power from nuclear processes was pure "moonshine," and that such research would lead to nothing useful. This article so irritated the Hungarian physicist Leó Szilárd (1898–1964) that he immediately visualized the nuclear chain reaction and invented the nuclear reactor before nuclear fission was discovered.

electromagnetic radiation and indicated that they were tiny particles, an idea that was definitely backed up by J. J. Thomson's work. Some of the particles would make it straight through, but some seemed to hit something hard and be absorbed. He noticed that the amount of absorption of the cathode rays was roughly proportional to the density of the material they were shot through. Moreover, the rays could make it through inches of air but were scattered by it, indicating that the air was composed of particles that were heavier than the cathode ray particles.

From those observations, Lenard made an unacceptable conclusion: The atoms, of which matter is composed, are made of almost entirely empty space. He intensified his assertion with a metaphor, saying that the volume occupied by a cubic meter of platinum was as empty as outer space. Within four years, Rutherford would come to agree with him.

THE DISCOVERY OF THE ATOMIC NUCLEUS

By 1906, Rutherford was still at McGill University in Montreal puzzling over Philipp Lenard's conjecture from 1903 concerning the void between atoms, and he was studying his newly discovered alpha particles. He was measuring the degree of deflection he could obtain using a strong magnetic field with alpha particles streaming through it. They were moving fast and were heavy, and to get a barely measurable deflection he had to use the most powerful magnet he could devise in the laboratory. His results were recorded on photographic film, showing where in space his beam of alphas landed after traversing the face of the magnet. He defined the beam using a narrow slit through a sheet of metal, and at one point he tried to improve the quality of the beam by putting a thin sheet of mica over part of the slit.

The mica was thin enough to allow alpha particles through, but the particles that came through the mica made an odd, blurred image on the film. As hard as it was to believe, the thin piece of mica was deflecting alpha particles through two degrees, and that was better than he could get using his best magnet. Rutherford made a calculation. To deflect alpha particles by two degrees would take an electrical field of 100 million volts per centimeter of mica. It was clear to him that the center of an atom had to be the source of very intense electrical forces. Alpha particle scattering required further study.

Back in Manchester in 1910, Rutherford set up his colleague Hans Geiger (1882–1945) and an undergraduate Ernest Marsden (1889–1970) to study this business of deflection of alpha particles through thin materials.

It would turn out to be a life-changing experience to be enshrined and known to physicists forevermore as "the Gold Foil Experiment."

Geiger and Marsden were going to try Rutherford's scattering experiment with a lot of other materials besides mica. They planned to try alu-

Rutherford's "Gold Foil" Experiment

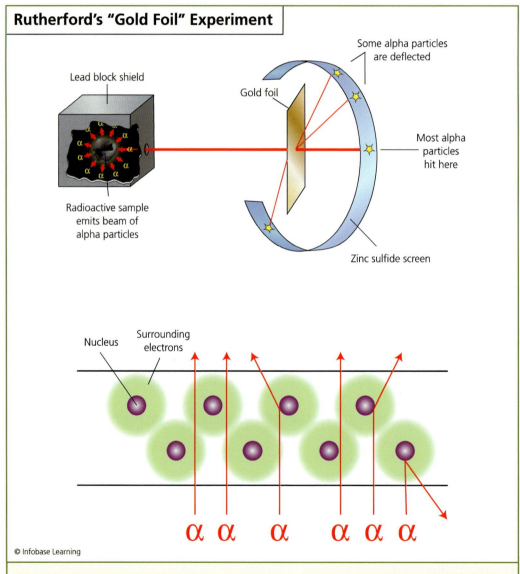

Geiger and Marsden, under Rutherford's supervision, directed alpha particles through a foil of gold to find the attenuating properties of a thin metal shield. To their astonishment, some particles were reflected backward out of the shield.

minum, silver, and platinum, all made thin enough for alpha particles to go through the samples, but first they would try gold because it was easiest to obtain in very thin samples. A vertical sheet of gold foil was set up. To count the alpha particles deflecting through the gold and note their positions they used a glass plate painted with zinc sulfide. It would glow or scintillate when hit with an alpha particle, and they would view it using an attached microscope with the lights turned off.

Next they needed a source of a beam of alpha particles. Radium was a convenient source, but it radiated alpha particles in all directions and they needed a tight beam. They built a special alpha source using a speck of radium at the end of a metal tube. The alpha particles would be absorbed in all directions in the tube except the direction leading right down the center. It seemed like a design that could not fail, but there was a problem. The tube was set so that it was aimed at the gold foil at a 45 degree angle. The pencil-thin beam was expected to deflect, going through the foil and coming out the other side in a spray four degrees wide, but there were alpha particles where there should be none, wide of the opening in the end of the alpha source tube. It appeared that the tube setup was faulty, and that alphas were somehow being emitted at odd angles by the tube. The two scientists tried to fix it. Nothing they tried seemed to work.

Rutherford wandered into the room to find out how it was going. Marsden reported unsatisfactory results. The beam was too wide, and they were detecting alpha particles scattered widely. Rutherford had an idea. He told Marsden to look for alpha particles in front of the foil, instead of in back of the foil, where the beam was supposed to emerge. Marsden slid a thick, lead shield between the viewing screen and the alpha tube to make sure he was not looking at stray alphas out of the source and put his eye to the microscope, mounted at a 90-degree angle on the front of the gold foil. Marsden was astonished at what he saw in the eyepiece. Instead of simply being deflected by as much as two degrees by going through the gold, the alpha particles were being deflected backward, by an astonishing 90 degrees or more. He met Rutherford on the steps leading to his private room and broke the news. Rutherford was overjoyed. A piece of gold 0.00002 inches (0.00006 cm) thick was deflecting alpha particles through an angle that would require one enormous magnet. As Lord Rutherford recalled the event later, "It was almost as incredible as if you fired a 15-inch shell at a piece of tissue paper and it came back and hit you."

Lenard's observation concerning the extreme lack of substance in matter had been absolutely correct, and Rutherford quickly adjusted his

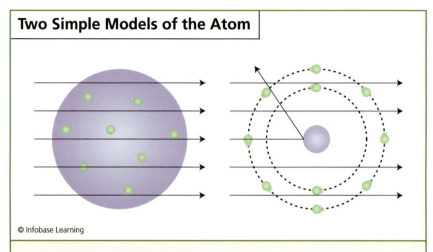

Two Simple Models of the Atom

© Infobase Learning

The left diagram shows Thomson's "plum pudding" model of the atom, consisting of protons evenly distributed through a round blob of negative charge. On the right is Rutherford's modified concept, a tiny, hard nut of protons at the center of orbiting electrons.

concepts to match it. The alpha particles trying to run through the gold atoms in the thin foil were like comets approaching a thin galaxy of stars in outer space. Get too close to a star, and the comet will whip around it and come back in nearly the same direction in a tight, parabolic trajectory. The astronomical analogy was obvious, and Lord Rutherford proceeded to model his atoms as Sunlike nuclei having planetlike electrons spinning around them in elliptical orbits. It made a certain poetic sense that matter would be composed of tiny solar systems. The universe in its tiniest form was the same as the universe in its largest form.

3 Monumental Theories

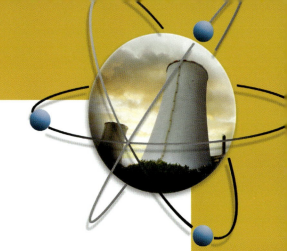

In the normal day-to-day world, quantities such as speed, mass, and distance are smooth and infinitely divisible. Actions are predictable and reproducible. In the submicroscopic world of quantum mechanics, quantities are jerky and make sudden jumps from one value to another. Actions seem governed by probability, and randomness prevails.

Imagine a block of *radioactive* cobalt metal, for example. It is an absolutely certain prediction that in 5.272 years the rate of radiation production in this sample of cobalt will decrease by half, as the metal block produces a continuous stream of *gamma rays*. Cut the block in half, and lay aside one of the pieces. The size and the radiation production of the sample are now smaller, but it is still a piece of cobalt and the radiation will still decrease by exactly half in 5.272 years. Now cut the remaining piece in half, and push one half aside. Cut this new piece in half. In theory, this block of cobalt can be cut in half a seemingly infinite number of times, but eventually the block of cobalt is so small it is just two atoms of cobalt stuck together. Cut that in half, and the block consists of one atom of cobalt. That atom cannot be cut in half, or at least if it were then the results would no longer be two, smaller pieces of cobalt. Cut an atom of cobalt in half, and all the characteristics of cobalt are lost in the process. The atom cannot be evenly divided, as the nucleus contains 27 *protons*. The atom is therefore the smallest quantity in which cobalt can exist, and the metal is not infinitely divisible. This smallest possible bit is

27

the quantum of *cobalt*. Every quantity of cobalt is some integer multiple of this quantum, but in the macroscopic world, or the world at a scale with which we are familiar, the digital graininess of this material is so small it is not noticeable. There is a threshold. Above the threshold is the continuum of classical physics. Below the threshold is the discontinuous region of probabilistic action and quantum mechanics. There is no way to predict when one atom of radioactive cobalt will decay and send off a burst of radiation. The best prediction is that within 5.272 years the odds are 50–50 that the atom will undergo *radioactive decay.*

Quantum mechanics would provide a firm, theoretical basis and explanation for how the energy release from nuclear decay could be achieved. At full force, quantum mechanics would predict the energy-producing qualities of nonexistent elements and the properties of previously unknown subatomic particles. Eventually, quantum mechanics would be used to back step time in a forensic study of the beginning of the universe. This chapter will reveal some of the most crucial beginnings of quantum mechanics as it was used to push forward an understanding of the atomic nucleus. As the physicist who invented it once said, "If quantum mechanics hasn't profoundly shocked you, you haven't understood it yet."

MAX PLANCK (1858–1947) AND THE ELEMENTARY QUANTUM OF ACTION

In 1879, most physics was experimental, and Max Karl Ernst Planck, Ph.D. in physics, found himself the only theorist in the Berlin Physical Society. A principle or finding that was first predicted mathematically was considered "spooky" in this German gathering, and purely theoretical studies had yet to achieve their full recognition as an essential part of the advancement of science. Still, in 1899, Planck won some funding from a consortium of electrical companies to discover how to derive the most light from lightbulbs using the least amount of power.

The problem to be solved boiled down to a central question: How does the intensity of the light emitted by a heated "black body," or a perfect light-absorber, depend on the frequency, or color, of the light and the temperature of the body? There had been unsuccessful explanations of black-body radiation, and many experimental results had been accumulated. Approaching the problem from a mathematical end, Planck found, to his despair, that the possible explanations for an energy-frequency relationship all seemed to center on the statistical mechanics studies of an Austrian physicist named Ludwig Boltzmann (1844–1906). Planck and many of his fellow physicists at the time held a strong aversion to these probabilistic, statistics-based interpretations. Deeply suspicious of the philosophical and physical implications of his own work, Planck sacrificed his convictions and derived an important equation:

$$E = h\nu$$

E is the energy of a *photon,* which is directly proportional to the frequency, *v,* of the photon. A constant, *h,* completes the conversion. The energy of light is dependent only on its frequency. The startling implication is that a light beam of a given frequency, *v,* cannot be divided down any lower than *hv.* Any further division, and the beam is no longer light. An intensity, or quantity, of light is simply an integer multiple of *hv,* but h is so small we do not normally notice any digital jump in brightness. The constant, *h,* became known as Planck's action quantum, or simply Planck's constant. The equation proved accurate in all experimental confirmations.

Although he received the Nobel Prize in physics in 1918 for having derived this formula, Max Planck had a difficult time believing in his own work. The concept of quantized energy destroyed his understanding of classical theory, and he was never able to accept it as a reality. Other stubborn, traditional physicists simply gave Planck's constant a value of zero and continued as if nothing had changed. Albert Einstein (1879–1955), a forward-thinking theoretical physicist in Germany, would explain the equation in his 1905 paper as an expression of light quanta, or photons, which were tiny, discrete particles of light. To the unending despair of both Einstein and Planck, quantum mechanics was born.

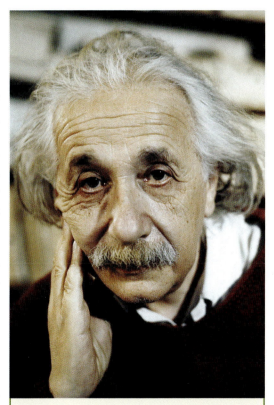

Although Einstein was disturbed by quantum mechanics, he gave birth to it with his Nobel Prize–winning theory of the photoelectric effect in 1905. *(Popperfoto/Getty Images)*

THE COPENHAGEN INTERPRETATION

Niels Bohr (1885–1962) was born in Copenhagen, Denmark, to Christian Bohr, a devout Lutheran and professor of physiology at the University of Copenhagen, and Ellen Adler Bohr, from a wealthy Jewish family engaged in Danish banking and politics. Young Niels and his brother, Harald, played soccer with a passion, but he had an even greater interest in science and studied under J. J. Thomson at Trinity College, Cambridge. He received his doctorate from the University of Copenhagen in 1911 with the thesis, "Studies in the Electron Theory of Metals." Bohr then moved to Manchester, England, for postdoctoral studies under Ernest Rutherford, applying himself to some problems with the new model of the atom.

Rutherford's proposed idea for the atomic structure had electrons literally orbiting a heavy nucleus, like planets orbiting a sun. It was an appealing model in that it matched well-known configurations, but the astronomical analogy was deeply flawed on the atomic level. Inter-electron interferences, in which like-charged particles get in each other's way in orbit, could be ignored by considering the simplest case: the hydrogen atom, with only one electron orbiting a simple nucleus consisting of a single positively charged proton. Even this case would not work. Technically, the electron in orbit was always accelerating, as its traveling direction had to constantly change as it maintained a circular path. According to Maxwell's equations, an accelerating electron would emit light. Emitting light would drain the electron of energy, and it would spiral down out of orbit and eventually crash into the nucleus, as well as making hydrogen glow continuously or until its atoms self-destructed. Hydrogen atoms had no tendency to self-destruct. They seemed to last forever, and there was no continuous glow from emitted light.

However, hydrogen could be made to glow under duress. In the great rush to explore vacuum tubes and high-voltage effects in the late 19th century, physicists had filled evacuated tubes with individual gases, such as argon, neon, xenon, and even nitrogen and hydrogen. They found that under the excitation of thousands of volts, gases would glow, with characteristic colors. An organized way to classify gases by excitation color was to direct the light through a slit, making a narrow beam, and then through a glass prism, separating it into a spectrum, using a device that had been around since 1859. The results were exciting and useful. A certain gas would throw narrow lines of consistently distinct colors on a scale representing the color spectrum of light.

There was no theory as to why gases such as hydrogen emitted light only in certain colors, but a Swiss mathematician, Johann Balmer (1825–98) came up with a formula that would predict, with amazing accuracy, the positions of the spectral lines from electrically excited hydrogen on a scale of light wavelength. The lines of hydrogen light were named the Balmer Series in his honor, and in 1885 there was much excitement over this finding but still no ideas as to why this formula worked.

In 1913, Bohr, working in Manchester, stared at Balmer's formula and realized something: To derive the correct light-wavelengths for hydrogen, numbers are plugged into the equation. Not just any numbers are inserted, but integers, such as 3, 4, 5, and so on. Immediately, the orbital structure of the hydrogen atom became clear to Bohr. The electron bound to the positively charged nucleus was normally in a basic orbit called a ground state. The ground state was the minimum energy a hydrogen electron could have and still be orbiting. Add energy to the atom, by establishing a high voltage across it in a tube, and the electron responds by jumping to a higher orbit. Add more energy, and the electron jumps to an even higher orbit. Remove the excitation, and the electron jumps back down to ground state. The term *orbit* had been misused. It was not literally a satellite-style orbit, but an energy state. It was when an electron jumps from one energy state to another that it actually experiences an acceleration and therefore radiates a Maxwellian particle of light into space. An electron can jump from its highest orbit to ground, from its highest orbit to a semi-highest orbit, or from a semi-highest orbit to ground, and each abrupt transition results in a different wavelength, or color, of light, as predicted by integers inserted into Balmer's formula. Each energy transition represents a quantum leap, instead of a continuous decay of an orbit, and this new model neatly explains Planck's finding that there is an indivisible quantum of light. One electron in one hydrogen atom, making one orbital transition produces one photon of a predictable color, and there is no way to divide that process down to make a smaller quantity of light.

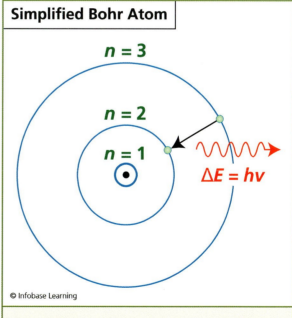

Simplified Bohr Atom

$n = 3$

$n = 2$

$n = 1$

$\Delta E = h\nu$

© Infobase Learning

In Niels Bohr's concept of atomic structure, electrons are in numbered, distinct orbits, each of a different energy level. When an electron does a sudden transition from an upper level to a lower level, the energy difference is radiated out of the atom as light.

Not only had Bohr explained why Balmer's equation seemed to work and affirmed Planck's findings, he had validated Einstein's energy-packet theory of light, and further experimental confirmation impressed physicists worldwide. He had invented quantum mechanics.

NIELS BOHR IMPOSES QUANTUM MECHANICS ON THE ATOMIC MODEL

In the 1920s, Bohr's quantum mechanics was widely accepted and embraced by the community of nuclear physicists, and he continued to refine and expand its implications for the atomic structure. Once hydrogen, the simplest possible case, was well defined by quantum theory, Bohr turned to the more difficult task of explaining atomic systems having numerous electrons in orbit.

COUNTERINTUITIVE ASPECTS OF QUANTUM MECHANICS

In 1801, Thomas Young (1773–1829), an English polymath with a doctorate in physics from the University of Göttingen, decided to prove that light was a wave phenomenon, such as sound. He set up an opaque screen, in which two vertical slits were cut and spaced about an inch apart. He directed a beam of light into the slits on one side of the screen, and on the other side he observed the resulting pattern of light on a piece of white paper.

The experiment was entirely successful, showing an interference pattern of alternating light and dark lines projected onto the paper. From this simple setup, there is no doubt that light travels as a wave. The wave front encounters the screen, where it is divided into two sub-waves, which recombine in space and then hit the paper with a predictable interference pattern. The experiment has been rerun innumerable times, using all known types of electromagnetic radiation and even beams of high-speed particles. In 1961, the experiment was run using a beam of electrons, and in 1989, it was run successfully using a single electron aimed at the screen between the two slits. The results are always the same.

In 1913, Niels Bohr confirmed the theories of Planck and Einstein, that light is not a wave but a particle, with his quantum theory of the orbital mechanics of

Bohr modeled the ground energy state of an electron as a sphere of hypothetical altitude above the central nucleus. These ground energies occur in integral steps, of course, and as the ground energy becomes higher, the sphere has a larger simulated diameter and can accommodate a larger number of electrons. It is possible for the first ground state to have as many as two electrons in it. Hydrogen has this ground state, but it contains only one electron, as there is only one positive charge in the hydrogen nucleus to counteract one negative charge. This ground state is only half filled. Helium had two protons in the nucleus and could accommodate two electrons, so its ground state is filled. Higher ground state energies, or shells, can accommodate eight electrons, 18 electrons, then 32 electrons, and so on as the shells grow larger around the nucleus.

With Bohr's orderly arrangement of electrons in an atom, the physical meaning of the periodic table of the elements, as invented in 1869 by

atomic electrons. Furthermore, Arthur Compton (1892–1962), head of the physics department at Washington University in St. Louis, proved in 1922 that photons lose energy when they crash into something solid and exchange momentum with electrons. Waves do not do that. Only solid particles exchange momentum and scatter like balls on a billiard table. From Compton's experiments, there is no doubt that light travels as particles. In the single-electron experiment of 1989, however, an electron, which is a particle, apparently divided in half and interfered with itself on the other side of the slit screen.

Scientists found, and still find, these experimental results profoundly baffling. Light can be a particle or light can be a wave, but it cannot be both. Set up both experiments, in tandem. First, divide the wave with the double slits, and then determine that the light coming out of one of the slits is a particle. When the light is detected as a particle, the interference pattern from the two slits disappears. The light can be determined to be a particle or a wave, but not both simultaneously, and the nature of the measurement determines what it shall be. It seems as if light knows how it is being perceived and adjusts its identity according to the experiment.

Niels Bohr summed up the physical meaning of these findings: "Nothing exists until it is measured." Quantum mechanics is considered counterintuitive, or outside normal perceptions of reality, for these and other reasons.

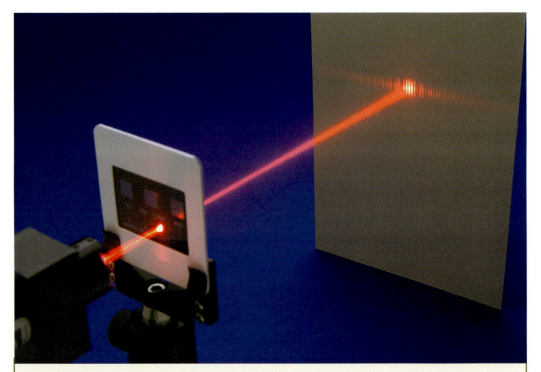

The double-slit experiment, showing a most vexing paradox of quantum mechanics: Light can be a particle or a wave, depending on how the experiment is set up. *(GIPhotoStock/Photo Researchers, Inc.)*

the Russian chemist Dmitry Mendeleev (1834–1907), became clear. Alkali metals, such as lithium, sodium, and potassium, are arranged in a column in the periodic table, as they all seem to have a similar chemical characteristic. They have this common characteristic because all elements in this column have a single electron in the outermost energy shell. It is the outer shell that determines an element's chemical interaction, and any lower ground state electrons do not contribute at all to compounding. The inert gases, such as helium, neon, argon, and krypton, all have a completely filled outer shell. With no unfilled electron stations in this uppermost ground state, there is no way one of these gases can bind with another element. Bohr went so far as to predict that when element number 72 was discovered, it would have four electrons in its outer shell and therefore behave chemically like zirconium.

4 Nuclear Fission Is Discovered

In the 1920s, theoretical physics seemed to flourish and move forward, while experimental physics was stalled and making no headway. Niels Bohr's creation of quantum mechanics led to new and deeper insights into forces and structures that were too small to be detected using the available experimental equipment. Excellent theoretical work could go only so far without experimental results to back it up, and there was ongoing work in England, France, and Germany to analyze the construction of the atomic nucleus as the new theories grew in acceptance and importance.

On June 3, 1920, Sir Ernest Rutherford gave the Bakerian Lecture at the Royal Society of London, on the successful transmutation of the nitrogen atom using alpha particles. Out of character, Rutherford diverted off the topic of his nitrogen experiment into speculations concerning the constitution of the nucleus. It was known by this time that atoms were composed of a fixed number of electrons clouding a nucleus that is composed of a like number of protons. The negative charges of the electrons perfectly cancel the equal positive charges of the nucleus, leaving an electrically neutral atom. The majority of the atomic weight is due to the protons, which are heavy particles, jammed closely together, making a dense nucleus. It was a workable model of the atom, and quantum mechanics would come to explain how chemical reactions work given no more detail than this. However, there was a serious problem. Hydrogen has one electron and one proton. Helium has two electrons and two protons. A helium atom should weigh exactly twice what a hydrogen atom weighs. It does not. A helium atom weighs four times what a hydrogen atom weighs. If hydrogen has an atomic weight of one, then helium is four. Moreover, nitrogen has seven protons, but an atomic weight of 14, and the disparity grows worse as the atoms grow heavier. Barium has 56 protons but weighs 138. Uranium has 92 protons and weighs 235 or 238, depending on which isotope of uranium is weighed.

In his now famous lecture, Rutherford proposed a solution to this puzzling aspect of nuclear structure. Nuclei above hydrogen are heavier than is explicable. There must be another particle at work in the nucleus. It is a particle with no measurable electrical charge, but it has all the weight of a proton. It is electrically neutral, and it should be called the *neutron*.

Such a particle would have interesting properties. Because it has no electrical charge, it would be free to go in and out of matter without being stopped by electron clouds covering atoms. It could not be contained by any solid walls, such as in a glass tube, or even by blocks of lead, and it would be free to enter the atomic cloud, penetrate cleanly to the center of the atom, and crash into the nucleus without being stopped. Having no charge, it would not leave an ionized trail as it flew through gas, liquid, or solid, and therefore it could not be detected with any known method of particle measurement. It was indeed an interesting particle for these reasons, but it was pure speculation. Nobody had ever seen even indirect evidence of a neutron beyond the observations of atomic weight.

In 1932, a research assistant of Rutherford's would finally find the neutron, in a skillful interpretation of experimental results. From his work, experiments with the newly found particle led to the discovery of nuclear fission, and from there experimental physics took the lead in the systematic development of nuclear power.

JAMES CHADWICK PROPOSES A NEUTRON

James Chadwick (1891–1974), son of a businessman from Cheshire, England, applied to the University of Manchester at age 16. He planned to major in mathematics, but as he stood in the queue for his entrance interview he realized too late that he was in the wrong line. He was too embarrassed to admit it and wound up in physics. His first year was miserable, as the physics classes were big and noisy, but when he heard a lecture by Ernest Rutherford he was converted to physics. He graduated in 1911 with a bachelor's degree in physics and went on to Cambridge, where he earned a master of science degree in physics in 1913. From there, he was awarded a scholarship to engage in nuclear research with Hans Geiger (1882–1945) at the University of Berlin, a valuable opportunity

Caught in Germany at the beginning of World War I, Chadwick was kept in a prisoner of war camp for nonmilitary aliens for the duration of the war. Released at the end of the war, he was hired by Rutherford to resume his research at the Cavendish Laboratory. His assignment was to look for the neutron, Rutherford's theoretical particle, and it would be a long search. He started by studying what others in the field were doing. They were bombarding light elements with alpha particles to see what would happen.

The alpha particle, as discovered and named by Rutherford, is an extremely heavy particle of great energy. It consists of two protons and two neutrons stuck together, and it is literally the nucleus of a helium atom. In the 1920s, it was only known to be heavy and positively charged. Hit a light element, such as boron or aluminum, with an alpha particle and the nucleus disintegrates, throwing off a burst of gamma rays and proton debris. Oddly, beryllium emitted a rash of gamma rays 10 times that of other elements bombarded with alphas, and there was no proton debris from a supposed nuclear destruction. No one knew why.

Irène Joliot-Curie (1879–1956) and her husband, Frédéric Joliot-Curie (1900–58), had reported detecting protons being knocked out of a sheet of paraffin by the gamma rays produced by the alpha-beryllium experiment. Chadwick was highly skeptical of the French findings. He believed their observations of the radiation were correct, but their explanation was questionable. They were saying that gamma rays from the beryllium were knocking protons out of solid wax. It was true that gamma rays of sufficient energy could deflect electrons, but protons are 1,836 times heavier than electrons. Saying that gamma rays were tossing protons into the ion chamber was like saying that a dump truck could be knocked into the oncoming lane by hitting it with a well-thrown baseball. Chadwick set up his own version of the Joliot-Curie experiment.

The protons were indeed knocked into the detector in Chadwick's setup. The only thing that could exchange momentum with a stationary proton and send it flying at high speed into the ion chamber would be a particle of the same weight, hitting it hard. There were no protons coming out of the beryllium, or they would have shown up in the ion chamber. Chadwick proposed a logical explanation. The particles coming out of the beryllium were neutrons. The neutrons made no impression on the radiation

The toolbox of James Chadwick (1891–1974), ca. 1932, which he used to discover the neutron. These are silver and aluminum foils of various thicknesses used as barriers to assess the strength of radiation, which he kept in a cigarette box. *(© SSPL/Image Works)*

detection apparatus simply because they were electrically neutral particles, but by hitting the paraffin a secondary effect could be seen, as the protons recoiled from billiard-ball collisions with the flying, invisible particles.

On February 17, 1932, after getting very little sleep over the previous week of intense work, James Chadwick published an announcement in the science journal *Nature* titled "Possible Existence of a Neutron," and the productive era of nuclear physics began.

LEÓ SZILÁRD THINKS OF THE SELF-SUSTAINING CHAIN REACTION

News of Chadwick's discovery traveled quickly, from England to Denmark to France to Germany, as experimentalists and theorists alike turned their rapt attention to the new particle and its interactions with the nucleus. To those working in theoretical quantum mechanics, the presence of a neutral particle in the nucleus made perfect sense. They had modeled the outer electron cloud as fixed stations of energy levels. As electrons were induced to jump from one level to another, the acceleration of the negative charge caused electromagnetic radiation in the form of light. The protons in the nucleus were obviously bound together by some strong force. It had to be stronger than the electromagnetic force that caused like charges to repel each other. They called it simply the strong nuclear force, and it has to depend on the neutrons. Alone, the protons lack enough force to hold themselves together.

The protons and neutrons making up the nucleus have their own abstract orbitals, as if they were swirling around in the tight, limited space at the center of the atom. Each orbital station in the nuclear orbit structure has an energy associated with it. The forces holding the nucleus together are millions of times more powerful than the forces holding electrons in orbit. Disturb the nuclear structure by knocking out a neutron, for example, and the nucleus has to reconfigure itself, with protons and neutrons jockeying for position and changing orbits. The severe change of energy status of a proton, with its positive electrical charge, causes a powerful electromagnetic pulse to radiate from the nucleus, in accordance with Maxwell's equations. Quantum mechanics thus explained the presence of gamma rays, the penetrating photon radiation produced in the beryllium experiment.

On Tuesday morning, September 12, 1933, Leó Szilárd (1898–1964), a brilliant physicist from Hungary, happened to be in London, lounging in the lobby of the Imperial Hotel and reading the *Times* newspaper. The headlines were all about nuclear science. "BREAKING DOWN THE ATOM," "TRANSFORMATION OF ELEMENTS," "THE NEUTRON," and halfway down the second column, a summary of a speech by Ernest Rutherford, "HOPE OF TRANSFORMING ANY ATOM." There was a scientific meeting going on involving all the top scientists in England, and Szilárd was acutely aware that he had not been invited. He started reading about Rutherford's speech:

What, Lord Rutherford asked in conclusion, were the prospects 20 or 30 years ahead? . . . We might in these processes obtain very much more energy than the protons supplied, but on the average we could not expect to obtain energy in this way. It was a very poor and inefficient way of producing energy, and anyone who looked for a source of power in the transformation of the atoms was talking moonshine.

Rutherford was saying that nuclear power on an industrial scale is impractical and not worth thinking about. Szilárd found such pronouncements bothersome. He tossed away the paper and wandered out onto the street, where he could think while walking. He was so put off by a scientist of such large reputation saying that something could not be done without having tried it, he tried to think of a counterargument.

He stopped at a traffic light on Southampton Row, at Russell Square, across from the British Museum in Bloomsbury. The light turned green, and just as he stepped off the curb to cross the street an idea flashed through his mind. Neutrons have no charge and are not constrained by the shielding effects of the electron or the proton. A neutron is free to hit the nucleus head on, if it is so directed. If an extra neutron wandered

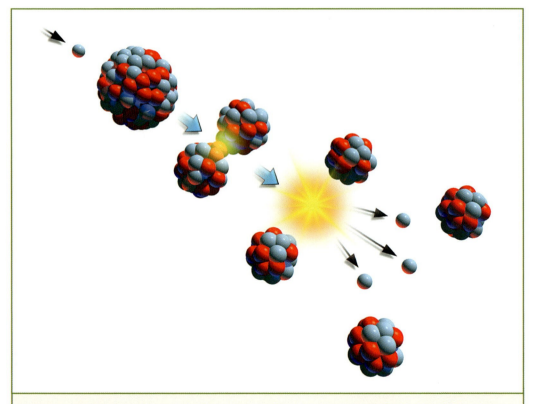

A diagram of the nuclear fission chain reaction, as first envisioned by Leó Szilárd in 1933. A free-traveling neutron is absorbed by a large atom, which becomes unstable and tears into two smaller atoms. Two or three stray neutrons are part of the fission debris. *(Andrea Danti, 2008, used under license from Shutterstock, Inc.)*

into and was absorbed by an already heavily laden nucleus, such as in one of the heavier elements, it could render it unstable. The unstable nucleus would then vibrate itself apart, rending into two nuclei. These two nuclei would almost surely weigh less together than the original large nucleus, and the weight deficit would express as pure energy.

Furthermore, suppose the destruction of the nucleus includes a single neutron scattered out of the debris. That neutron could then bounce around until it hit another overloaded nucleus, and it would cause another nuclear breakdown. Some neutrons would fail to cause subsequent breakdowns, just because there was only a finite probability of one hitting an adjacent nucleus. What if, instead of one free neutron from the breakdown, there were two? If there were as many as two individual neutrons in the breakdown debris, then the process could be self-sustaining. It would be a *chain reaction,* in which energy was released by nuclear disintegrations in quantities millions of times greater than any chemical reaction.

By the time he reached the other side of the street, Szilárd had outlined the nuclear power process. If such an element exists that will tear apart under neutron bombardment and will release free neutrons in excess of one per disintegration, then Lord Rutherford was wrong. Nuclear power on an industrial scale would be possible. He spent the rest of the day thinking of exploring the solar system and beyond with nuclear-powered rockets and of building weapons based on the severely concentrated energy of nuclear reactions. In 1934, his patent application for a nuclear power reactor was not granted, but the application document was assigned to the British Admiralty for security reasons. Leó Szilárd would live to see his daydreams realized.

THE REMOTE COLLABORATION OF OTTO HAHN (1879–1968) AND LISE MEITNER (1878–1968)

Otto Hahn was born in Frankfurt, Germany, the son of a prosperous glazier and property owner. He led a sheltered childhood, and at 15 he became interested in chemistry, performing experiments in the laundry room of the family home. Although his father wanted him to study architecture, Otto convinced him that industrial chemistry would be a better occupation. He began his studies in chemistry and mineralogy at the University of Marburg in 1879 and received his doctorate in chemistry in 1901. After completing a year of required military service, he returned to Marburg to work as a research assistant. In 1904, he took a job at University College London in England, working in the new field of radiochemistry. He moved back to Germany in 1906 and by 1910 was head of the radioactivity department of the new Kaiser Wilhelm Institute of Chemistry in Berlin.

In 1907, Hahn met Dr. Lise Meitner at a University of Berlin physics colloquium. She was from Vienna, Austria, and had already published papers on alpha and beta radiation. They both needed collaborators, and as a physicist and a radiochemist they would make a strong team. They quickly became friends and worked closely together for 30 years.

When Chadwick announced the discovery of the neutron in 1932, radiochemistry assumed a heightened importance. It was suspected that neutron capture by a nucleus would cause transmutation to another element, one atom at a time, but neutron sources were weak and the effect could be small. A good polonium-210 source was desirable but unavailable to most labs. Researchers used radium salt, an alpha particle emitter and a workable substitute for polonium, mixed with beryllium powder to make neutrons, using the effect discovered by Chadwick. There was no way to detect a transmutation except by exceedingly careful chemistry that would detect tiny amounts of a new element mixed in with the original sample. The amount of transmutation to be detected could be as small as hundreds of atoms. Otto Hahn was the world's leading radiochemist, and he used fractional crystallization as a method of finding minute contaminants in a sample. It is a method that was pioneered by Marie Curie, and it uses the fact that different substances dissolved in water crystallize out of supersaturated solution at different temperatures.

In 1938, Hahn and Meitner were getting some interesting chemical results of bombarding uranium with neutrons, but in July the political conditions in Germany for Meitner, who was Jewish, were becoming dangerous, and she had to drop everything and escape to Holland. The unsafe atmosphere for Jewish people was quickly becoming critical, and Germany seemed to be preparing for war. Hahn's work in radiochemistry was superb, but he needed Meitner's knowledge of nuclear physics to help interpret experimental results. With Meitner in exile, the two scientists kept in touch by mail and with one secret meeting in Copenhagen, Denmark.

Hahn repeatedly performed chemical analysis of the neutron-exposed uranium with very puzzling results. It was obvious that the neutron exposure resulted in new, radioactive elements in the uranium sample, as the radiation-counting instrument showed increased radioactivity, but the product of this transmutation was not clear. Hahn guessed it was a new, previously undiscovered isotope of radium. To detect small contaminants of radium, Hahn used barium as a carrier for the fractional crystallization. Meitner did not believe it was radium, as it would require a double alpha disintegration to go from uranium down to radium. To Hahn's amazement, all his tests for radium were negative, with nothing showing up in the barium carrier. In a burst of insight, Meitner figured out the problem, and she and Hahn exchanged excited letters on December 21, 1938. Hahn had not detected any radium in his barium because there was no radium. The product of the neutrons hitting the uranium was radioactive barium.

A uranium atom is slightly less than twice the mass of a barium atom. The uranium nuclei had split roughly in half, resulting in one barium atom and one krypton atom per disintegration. Krypton is an inert gas, undetectable by chemical means. The barium atoms in Hahn's experiment were neutron-heavy and therefore unstable and subject to radioactive decay, as was detected by his radiation counter. The next day, on

A reconstruction of the setup used by Otto Hahn and Lise Meitner to discover nuclear fission in 1938 *(Foto Deutsches Museum)*

LISE MEITNER: A REFUGEE SCIENTIST

Born in Vienna, Austria, on November 17, 1878, Elise Meitner was the third of eight children in a prosperous Jewish family living in the Leopoldstadt suburb of Vienna. Slight of figure, shy, and a formidable scientist, she was the second woman to earn a Ph.D. in physics at the University of Vienna. For reasons unknown, Elise shortened her first name to Lise and her birthday to November 7. Her teaming with Otto Hahn in 1907 would result in one of the most fortunate collaborations in the history of nuclear science, but its culmination in the discovery of fission would occur with her in exile and unable to share the credit.

In 1932, with Chadwick's discovery of the neutron, an unofficial scientific race began. Dr. Meitner rejoined the person who had invented applied radiochemistry, Dr. Hahn, for an investigation of the effects of neutrons on uranium. Also in the competition were Ernest Rutherford in Britain, Irène Joliot-Curie in France, and Enrico Fermi (1901–54) in Italy.

The next year, Adolf Hitler was named chancellor of Germany, and life immediately became very difficult for German Jews. It became illegal for them to work in German technical institutes or universities, and most, including Leó Szilárd, were forced to leave the country. Lise Meitner got by on a technicality. She was Austrian, not German, and she kept working at the Institute for Chemistry in Berlin as if she were immune to the Nazi mind-set. She and Otto Hahn were approaching an important

December 22, Hahn submitted his paper cautiously titled, "Concerning the existence of alkaline earth metals resulting from the neutron irradiation of uranium," and the discovery of nuclear fission was announced. Two months later Hahn wrote a second paper predicting the liberation of at least two neutrons during the fission process. With the discovery of fission in uranium, Szilárd's nuclear power reactor was not only possible, it was almost inevitable.

THE RACE IS ON

The fission effect in uranium was confirmed in laboratories from England to Japan, and its profound implications were well understood by all engaged in nuclear physics. In 1939, nuclear fission was seen as having two possible uses.

discovery, methodically and with patience. By 1938, the team had identified the half-lives of 10 unknown radioactive substances in uranium bombarded by neutrons.

On March 14, the political climate changed for the worse. Austria was annexed by Germany, and Lise Meitner became a German citizen. She had to leave quickly and surreptitiously, before she was arrested for being a Jew. Friends smuggled her across the border to Holland, and she made it out with enough money to buy lunch and with no possessions. She wound up working in a nuclear physics laboratory in Stockholm, Sweden, but she kept up with Hahn back at the laboratory in Germany by mail.

They met clandestinely in Copenhagen in November, planning a new set of neutron bombardment experiments, and the letters kept flowing. Hahn steadfastly believed that nuclear fission was impossible. In late December, Meitner convinced him that he had split uranium into two large fragments, one of which was barium, and with that realization history was made. Unfortunately, it was illegal for Otto Hahn to put Lise Meitner's name on the paper describing the chemical findings as coinvestigator because of her religious affiliation. Meitner and her nephew, the nuclear physicist Otto Robert Frisch (1904–79), published a paper two months later giving the physical explanation of Hahn's discovery and naming the effect "nuclear fission."

Lise Meitner died in Cambridge, England, in 1968 after a life of dedicated work in nuclear physics. Element 109, meitnerium, was named in her honor in 1997. The most stable known isotope of meitnerium has a half-life of 1.1 seconds.

The first application is a controllable source of constant power. Neutrons, which are necessary to cause the chain reaction effect, are ejected from a fissioning nucleus at high speed. They can be slowed down to a crawl by repeated collisions with surrounding light nuclei, and at the low speed the free neutrons can be captured by uranium nuclei, causing further fission. The act of being slowed down imparts energy to the light nuclei, and this material, known as the moderator, is then used as a heat-transfer medium. The process can be controlled easily, by balancing the number of neutrons being produced with the number of neutrons necessary to keep the chain reaction stable.

The second possible application is a superbomb. Neutrons produce fission at slow, *thermal speed,* but there is also fast fission from neutrons at the extreme high end of the speed range. Start a chain reaction in a large enough mass of pure uranium, with no intervening moderator material, and the fission will run out of control. It happens with such speed and is such a huge explosion that it can destroy an entire city. The military purpose of this application was both obvious and terrifying, and it was also seen immediately.

In February 1940, Otto Frisch, Lise Meitner's nephew, and Rudolph Peierls (1907–95) drafted a memorandum to the British Committee on the Scientific Survey of Air Defense, titled, "On the Construction of a 'Super-bomb' Based on a Nuclear Chain Reaction in Uranium." Similar memoranda were written in Germany, Japan, and the Soviet Union at about the same time.

Scientists in every country considering the military uses of nuclear fission realized that there was a problem with applying this effect. The useful fission seemed to occur in only one isotope of uranium, U-235, and it was only a small component of naturally occurring uranium. To build a bomb, the 235 isotope had to be nearly pure, and separating U-235 from *U-238* was no simple process. In the long run, only the United States had the industrial volume, ability, and materials to achieve this huge task. On top of that, the United States had inherited a large portion of the nuclear physics talent from Europe, as it fled the threatening fascist governments. The United States, which had contributed little to the field of nuclear research, became the world's center for it during World War II, and science, technology, and international affairs would be changed forever.

5 A Gathering of Nuclear Scientists in the United States

The year 1939 was a critical turning point not only in the state of the world but also in the development of nuclear power. A worldwide economic depression had been in effect for 10 years, and the lack of investment resources did not encourage the development of an entirely new energy source. Although it was scientifically possible to achieve a self-sustaining nuclear reactor that would generate thermal power, it was financially impossible. There were innumerable engineering details to be worked out, and there were still scientific unknowns and blank spots in the theories.

The biggest problem was dealing with the very low concentration of a usable isotope of uranium, U-235, mixed in with an unusable isotope, U-238. As it occurs in nature, uranium contains only 0.7 percent of the fissionable isotope that could be used for power generation. Isotope separation, in which the concentration of U-235 in natural uranium could be improved, had been demonstrated only as laboratory setups involving countable numbers of atoms. To make practical nuclear fuel would require an industrial effort of such enormous scale it was out of the question for normal commerce.

This chapter focuses on the factors that made 1939 a pivotal year in the development of nuclear power. The components of this historical convergence include an ominous but ultimately false fear of atomic bomb construction in Germany, the discovery in Italy of neutron interactions at low energies, a fortuitous transfer of nuclear physics expertise from Europe to the United States, and the development of an infrastructure for high-energy particle physics in California. In the middle of this frenzied race for nuclear fission, a second mode of nuclear energy production, fusion, was theorized into being. Although it seemed out of place, this theory completed the scientific understanding of energy from the nucleus, and its importance is undeniable.

THE FEARED THREAT OF A GERMAN ATOMIC BOMB

By 1939, it was apparent that the government of Germany was rearmed and preparing to reclaim territories on the eastern and western borders that had been lost in World War I. An alliance with Italy and Japan caused concern in Europe and the United States, as another large-scale conflict seemed inevitable.

To compound these concerns, power-producing fission had been discovered in Germany, the country that was now making aggressive moves. It was a logical conclusion that the Germans would be pursuing the development of nuclear weapons, as they were scientifically able to do with a strong motive and a well-developed industrial economy. The use of nuclear armaments, with their extreme power concentrated into small singular weapons, would obviously shorten a major conflict and reduce its cost. These considerations were not lost on the major governments in Europe, and defensive measures were implemented quickly as a result. England rushed its Home Chain radar early-warning system into operation, and France shored up its Maginot Line buried defense system on the northern frontier.

The Germans did start several nuclear weapons programs in secret as war clouds were forming, but no such weapon was used in the European theater of World War II. As the smoke dissipated in 1945 after the German surrender, it became clear that there had been nothing to worry about after all. The Germans had not even achieved a self-sustaining nuclear chain reaction, much less a nuclear explosive. The reasons for this failure are interesting.

The German atomic bomb program began early, just months after the discovery of fission in uranium. It was given the name Uranverein, or the "Uranium Club." There were misgivings, even at this earliest starting point. Carl Friedrich von Weizsäcker (1912–2007), one of the German participants in the effort, remembers the feeling:

To a person finding himself at the beginning of an era, its simple fundamental structures may become visible like a distant landscape in the flash of a single stroke of lightning. But the path toward them in the dark is long and confusing.

Weizsäcker and others worried that wars waged with atomic bombs could not be won, as both sides would be wiped out. "But the atom bomb exists," he said. "It exists in the minds of some men."

The first task facing the Uranium Club was to produce a self-sustaining nuclear reaction, and for that purpose a neutron moderator, or a material to slow neutrons to fission speed, would be needed. An obvious material was graphite, because carbon would not capture neutrons and remove them from the fission process. A low neutron capture probability was critical at this beginning point, because of the very small component of usable U-235 in mined uranium. The sustained nuclear fission in a chain reaction experiment using natural uranium would barely work, and it would be sensitive to neutron loss. The loss of a tiny percentage of fission neutrons to nonproductive absorption would shut the reactor down. This meant that graphite was unusable, because European industrial graphite contained trace amounts of neutron-absorbing impurities, such as boron. For the same reason, pure distilled water would be unusable. The hydrogen in water has a slight tendency to capture neutrons. A solution was to use *heavy water,* or deuterium-oxide. *Deuterium* is a hydrogen atom with a neutron precaptured, and it is unlikely to capture another. It was decided to build the German nuclear reactor using heavy water as a moderator. This decision would slow down the development of the German reactor considerably, as heavy water rarely occurs in nature, and it must be separated from natural water at great expense and effort.

The second reason for the loss of speed in the German atomic bomb development was that the entire Uranium Club was called to military service four months after it formed. The club was reformed on September 1, 1939, the day World War II officially started, and, under the direction of the Army Ordnance Office, research into the feasibility of nuclear weapons was conducted with moderate funding. The third reason for the loss of speed occurred on February 26, 1942, when a critical meeting was held by the Research Council to report on the technical progress and make recommendations for further work. All the top people in the German government and military were invited, but unfortunately included with the invitation was a copy of the lunch menu. Trying to impress the government officials with the depth of their research expertise, the scientists had planned an experimental lunch. The food was to consist of several types of vitamin-enriched morsels, all fried in "synthetic lard." All the invited officials found something else they had to do that day, and the German atomic bomb program sustained a crippling delay.

Perhaps the most important reason for the slowdown in the German nuclear development program was the emigration of some of the finest theoretical and experimental physicists in the world. Newly implemented laws under the Nazi and Fascist governments of Germany and Italy cleared the universities of all research and academic faculty of Jewish heritage and even visiting Jewish professors from Hungary were forced to leave. They spread west, first to Great Britain and then to the United States.

THE INTERESTING EFFECTS OF NEUTRONS AT LOW SPEEDS

The U.S. patent for a neutronic reactor has on it the names of two immigrants: Leó Szilárd from Hungary and Enrico Fermi (1901–54), a leading nuclear physicist from Italy. It was originally Szilárd's fanciful idea to create a sustained chain reaction of nuclear fission even before fission was discovered, but it was Fermi's genius that reduced the concept to a working, physical machine, capable of releasing power from the atomic nucleus. In doing so, he originated the discipline of nuclear engineering. Although the patent application was made in 1944, as World War II raged on, the document remained secret for 11 years, and the patent was not awarded until 1955, a year after Fermi died of stomach cancer. The nuclear reactor is considered one of the most important inventions in U.S. history. It won Fermi a place in the National Inventors Hall of Fame. In a recent poll by *Time* magazine, he was listed among the top 20 scientists of the 20th century.

The invention of the nuclear reactor was not necessarily Fermi's greatest accomplishment. In 1933, before he came to the United States, Fermi had formulated a theory of the beta decay of elements. To electromagnetic force, gravitational force, and the strong nuclear force that holds the nucleus together, he added a weak nuclear force, responsible for the breakdown of neutrons and protons and the release of electrons and positrons as radiation, or negative and positive beta particles.

Enrico Fermi was born on September 29, 1901, in Rome, Italy, to Alberto, a chief inspector of the Ministry of Communications in Rome, and Ida de Gattis. His immersion into the study of physics probably began at 14, when his older brother, Giulio, died unexpectedly during surgery for a throat abscess. The boys had been very close. The death hit young Enrico hard, and he sought a diversion. Soon after, he was browsing in the stalls of the Campo dei Fiori in Rome and discovered two volumes of *Elementary Mathematical Physics,* written in 1840 by a Jesuit physicist. He used his entire allowance to purchase the books and read them cover to cover multiple times. A friend of his father loaned him many books on physics and mathematics, and he studied each of them thoroughly. By the time he graduated from high school a year early, he had decided to dedicate himself exclusively to the study of physics.

Fermi was accepted at the Scuola Normale Superiore in Pisa, where he earned his undergraduate and doctoral degrees in physics. His entrance essay was considered exemplary and of Ph.D. thesis quality, and in college he became a great propagandist for quantum mechanics. In 1924, while working on his doctorate, he spent a semester in Göttingen, Germany, working with Werner Heisenberg (1901–76), the notable theorist in quantum mechanics, finding the philosophical, nebulous bend of this branch of physics hard to swallow. Fermi developed a style demanding concreteness and rigorous simplicity, inclined toward physical phenomena that could be confirmed by direct, unambiguous experimentation. He became a rare specimen of physicist, one who straddles the worlds of theoretical and experimental science with perfect balance. As a fellow physicist once remarked, "He was simply unable to let things be foggy. Since they always are, this kept him pretty active."

From studies in Germany, Fermi plunged into a professorship in physics at the University of Rome–La Sapienza in 1926. This would prove to be a challenging position, as Italy had a poor reputation in the physics community, and facilities and funding were at a subsistence level. Stepping boldly into the job, Fermi selected a competent team, soon nicknamed the "Via Panisperna boys." The men in his research group counter-nicknamed Fermi "the Pope." By 1933, Fermi had completed a detailed, quantitative study of beta decay in radioactive materials, complete with a fundamental theory of beta radiation. His paper describing this work, "Tentativo di una Teoria dei Raggi β" (An attempted theory of beta rays), was rejected by the journal *Nature* as being too removed from physical reality, so he published it in an Italian journal, *Ricerca Scientifica* (Scientific research). It would be six years before the editors at *Nature* realized their profound mistake and finally published the paper in English.

At the age of 33, Fermi and his team then engaged in one of the most important research efforts in the history of nuclear power. Using hand-built *Geiger counters* and neutron sources, they measured the interactions of neutrons with almost every element on the periodic table. One result of neutron bombardment to be measured was *activation,* in which a material absorbs a neutron and becomes radioactive. Some elements are very sensitive to activation, and some are not. When measuring the activation of silver, by accident the team found a puzzling effect. The intensity of the activation apparently depended on where in the laboratory the experiment was conducted. Most of the laboratory benches were topped with fine, Italian marble, but one bench had a wooden top, and this bench seemed

NIELS BOHR: THE LAST OF THE REFUGEES

Niels Bohr, the Danish physicist who founded quantum mechanics, almost waited too long to escape from Europe during World War II. The German Army overran and occupied Denmark in April 1940, but Bohr felt fairly safe in spite of his Jewish heritage. He was head of the Institute of Theoretical Physics at the University of Copenhagen, and it would have seemed unusually severe if the Germans had arrested him. Denmark essentially collaborated with the German Army, maintaining an uneasy peace. However, in 1941, Bohr was visited by Werner Heisenberg from Berlin. They had a long walk and a private conversation that left Bohr with a feeling of dread that scientists in Germany were working on an atomic bomb. Although he would deny it after the war, Heisenberg seemed to hint at atomic weapons research back in Berlin, while probing Bohr for indications as to what the British and Americans were up to.

By 1943, the situation in Denmark had worsened, and on September 28 Bohr learned from the Swedish ambassador that he was to be arrested and deported to Germany within three days. Wasting no time, Bohr and his family walked through Copenhagen to the seaside and hid until nightfall in a gardener's shed. A motorboat then ran them out to a fishing boat, which avoided minefields and German patrols and took them across the Oresund Channel to Limhamn, Sweden.

Sweden was relatively safe, but it was crawling with German agents, and there was real fear that Bohr would be assassinated now that he had escaped German control in Denmark. He wanted to get to Britain, at least, where he could warn the Allies of an impending German nuclear weapon. He was flown out in a British Mosquito to have miraculous qualities. The activation was much more intense when the experiment was performed on the wooden table.

It was a mystery worth further study. Fermi set up an experiment with a carefully machined piece of lead separating the neutron source and the silver target, but at the last moment, on a vague hunch, he substituted a scrap sheet of paraffin for the lead. The activation level increased dramatically over all previous experiments, and Fermi immediately knew what was taking place. The neutrons, barreling out of the neutron source at high speed, were slowed down to a crawl by elastic collisions with the hydrogen nuclei in the paraffin. Neutrons running slowly had more time to interact with the silver nuclei as they passed by, increasing the probability of being

twin-engine fighter-bomber, stripped of armaments but equipped with a compartment for a person to ride in a prone position where the bombs were normally kept. Bohr was strapped into a flight suit, with a parachute, a flight helmet with audio hookup, oxygen mask, and a handful of flares. They would be flying high, above 20,000 feet (6,100 m), to avoid German antiaircraft guns in Norway, and if they were caught by a fighter plane they were to open the bomb bay and drop Bohr into the ocean, in which case his flares would come in handy.

Unfortunately, Bohr had an unusually large head, and the standard issue flight helmet did not fit properly. He did not hear the pilot through the built-in headphones when he was told to turn on his oxygen after the plane made high altitude, and he passed out somewhere over Norway. The pilot could tell that something was wrong when he could get no verbal response from Bohr, and as soon as they were clear of Norway he dropped altitude and flew low over the North Sea. When they landed in England, Bohr was in fine shape, commenting that he had slept well during the flight.

From England, Bohr was flown to the United States, where he was taken to the top-secret atomic bomb laboratory in Los Alamos, New Mexico. Here he would add guidance, encouragement, and assistance to the theoretical work, as an expert in quantum mechanics. "An expert," he commented, "is a person who had made all the mistakes that can be made in a very narrow field." Upon seeing the extent of the American operation at Los Alamos, he was deeply impressed. Nothing of this magnitude had seemed possible anywhere in Europe. Although welcomed as a revered elder statesman of nuclear physics at the laboratory, he later confided to a friend, "They didn't need my help in making the atom bomb." They seemed to have it well in hand.

captured by the target. In the original experiments, neutrons bouncing off the wooden table had been slowed down, again by collisions with hydrogen nuclei in the wood. Hitting the heavy marble table top, the neutrons had not been slowed noticeably. To varying extents, the probability of neutron interaction would be increased as the speed of the particles was decreased, and this effect would apply to both absorption and fission. Fermi won the Nobel Prize in physics in 1938 for this discovery. Although he came very close to discovering fission, the Nobel Prize for that finding would go to Otto Hahn after World War II had ended.

At the subatomic level, all interactions of matter with matter are probabilistic in nature. If a freely traveling neutron flies close to a standing uranium atom, it does not necessarily do anything with the uranium, but there is a probability that it will be captured by its nucleus. The magnitude of the interaction probability depends entirely on the speed of the neutron as it passes. Although an interaction can occur at any speed, it seems that the slower a neutron is traveling, the higher is its probability of interaction.

Neutrons set free in uranium fission events are most likely traveling with an energy of about 1 *MeV*. (Neutron speed is expressed as neutron energy, which is always expressed in electron volts. An MeV is a million electron volts.) There is a probability that a fast neutron can produce an additional fission by hitting a nearby uranium nucleus, but it is a low chance. Slow the neutron down to thermal speed, or 0.025 *eV* (electron volts), and the probability of fission increases 1,000-fold. The term *thermal speed* means the speed at which air molecules normally move at room temperature. To consider building a machine that will operate as a nuclear fission reactor using natural uranium, as it is mined, then all favorable probabilities must be maximized. All probabilities unfavorable to fission, such as unproductive neutron absorption or leakage, must be minimized.

In his experiments on a wooden table and with paraffin wax, Fermi had found that if a high-speed neutron hits a hydrogen atom at room temperature, then the neutron and the hydrogen nucleus exchange momentum. The neutron slows to thermal speed and the hydrogen nucleus, which weighs about the same as a neutron, takes off at the speed of the original incoming neutron. This exchange between the energetic neutron and the room temperature hydrogen nucleus, or proton, would prove very important, as it is the mechanism by which the energy of fission can be transferred to a working fluid and exploited as power. Use water as the working fluid in a reactor, and the fast neutrons slowing down in it make steam.

Fermi, a Roman Catholic, had married Laura Capon, the daughter of a Jewish officer in the Italian navy, and he felt that anti-Semitic laws being enforced by the Fascist government of Italy were threatening his family. He took his wife and children to Stockholm, Sweden, to accept the prestigious Nobel Prize, and they never returned to Italy, slipping away and shipping instead to New York City for a new life in the United States. The United States gained a Nobel laureate, and Europe lost one. Fermi began work at Columbia University upon his arrival, and in 1942 he transferred his work to the University of Chicago. At this carefully selected location in the Midwest, under strictest secrecy, Fermi and his team of scientists built the first working nuclear reactor, Chicago *Pile* 1, and physics and the world would never be quite the same.

AN EXODUS FROM EUROPE

In the 1930s, Germany and Europe in general suffered a depletion of a resource that they had nurtured for decades. Starting in 1932, laws concerning the employment of people of Jewish heritage in universities and research organizations began to have serious effects. Germany began to lose the core of its valuable cache of theoretical and experimental scientists as they packed up and left the country. It was an inopportune time to have the number of available nuclear physicists reduced.

Hans Bethe (1906–2005) was a loyal German, reared in a Christian household, who became known as one of the few scientists who produced significant work for 60 years. He lost his job at the University of Tübingen in 1933 because his mother was Jewish. He then moved to the United States, joined the faculty of Cornell University, in New York, and became head of the theoretical division at the *Los Alamos* Laboratory during the atomic bomb project of World War II. Bethe calculated the critical mass of the weapons and did theoretical work on the implosion method used in the *plutonium*-based bombs.

Edward Teller (1908–2003) was a Hungarian Jew who moved to Germany for his education in chemical engineering and nuclear physics, earning a bachelor's degree at the University of Karlsruhe and a Ph.D. studying under Werner Heisenberg at the University of Leipzig. His job at the University of Göttingen was cut short in 1933, and he was invited to become a professor of physics at George Washington University in Washington, D.C. He joined the atomic bomb team at Los Alamos, New Mexico, during the war and went on to develop the *hydrogen bomb* and to be one of the founders of the Lawrence Livermore National Laboratory in California.

Another Hungarian Jew who immigrated to Germany to escape communism was Eugene "E. P." Wigner (1902–95). Wigner is the originator of most nuclear reactor theory as it is now practiced, and he won the Nobel Prize in physics in 1963 for his theories of symmetry in quantum mechanics. He studied at the Technische Hochschule in Berlin and worked at the University of Göttingen before he left Germany in 1930, seeing a deterioration of his fortunes as the Nazi regime coalesced. He was hired by Princeton University in New Jersey, and during the war he was named director of research and development at the Clinton Laboratory in *Oak Ridge,* Tennessee. His colleagues considered him the intellectual equal of Albert Einstein, and he was instrumental in convincing the U.S. government to begin the atomic bomb project.

These are only a few examples of the European scientists who fled to the United States because of repressive government policies. The United States had not been on the leading edge of nuclear research, but it would quickly become the world's center of it as these refugee scientists converged from Europe. In Britain, scientists had conducted research on a dignified, noncommercial scale, trying not to be extravagant or make outrageous predictions. The scientists in the United States may have been a second tier behind the Europeans, but they would not be held back by fear of extravagance. In the United States during World War II, a new type of science would be created. It would be science on a large scale, or "big science," with a direct connection to engineering and industrial processes. Billions of dollars would be diverted into some very risky research. The expatriate Europeans would be given anything they needed to continue their quest for the energy in the atomic nucleus. The large outpouring of effort and funding under war priorities would give nuclear power a push that would not have occurred under peaceful conditions with research conducted on an academic level.

PRELIMINARY NUCLEAR RESEARCH IN THE UNITED STATES

Nuclear physics research efforts in the United States were not entirely asleep before 1939. Although research expenditures for fundamental science were chronically short by today's standards, there was never a shortage of curiosity and a need to push physics into unknown territory. An area where the United States made bold progress during the Great Depression years was in particle acceleration. Charged subatomic particles, such as protons or electrons, can be accelerated electrically from rest energy up to great speeds, simulating radioactive decay products blasting free of nuclei but on a larger scale. The availability of a large flux of energetic charged particles made it possible to do element transmutation or nuclear disintegration on a scale order of magnitude bigger than had been accomplished using radioactive sources. Using a large enough accelerator, measurable quantities of isotopes could be manufactured in a laboratory, or enhanced resolution in nuclear probing was available. This class of machine came to be known as the atom smasher. The master of atom smasher design was Ernest Lawrence, of the University of California.

Ernest Lawrence (1901–58) was unique in the field of Nobel Prize–winning physics, in that his entire education was in the United States, without the degree from Europe that was considered necessary at the time. He started his college work at the University of South Dakota, transferred to the University of Minnesota, and received his bachelor's degree in 1922. He earned a master's in physics in 1923 and went to Yale University in Connecticut for a Ph.D. in physics in 1925. His most important contributions to nuclear research and development were two-fold. He invented the *cyclotron* atom smasher and the calutron large-scale isotope separator. He won the Nobel Prize in physics in 1939 for his work with the cyclotron.

Lawrence built his first cyclotron in 1929 at the University of California, Berkeley, using available materials. He called it his "proton merry-go-round" because it constrained charged particles to spin around in a tight spiral under a magnetic field as they gained power. The machine cost about $25 to build, the equivalent of about $313 today, and it was only five inches (13 cm) around, but it proved his point. Although the concept of the cyclotron probably occurred several places at about the same time, Lawrence dug in and built one. Leó Szilárd patented a cyclotron in Europe, but it takes more than a patent to break up nuclei. In 1934, Lawrence obtained a patent for the cyclotron, and by 1936 he had taken over the Civil Engineering Testing Laboratory, renamed it the Radiation Laboratory, and filled it with a 37-inch (94-cm) cyclotron capable of accelerating alpha particles to 16 MeV. He used it to create the first artificial element, technetium, a substance that is in the middle of the periodic table but does not exist in the crust of the Earth.

In 1939, Lawrence completed his 60-inch (150-cm) cyclotron, just in time to participate in the pivotal year in nuclear physics. It was a colossal

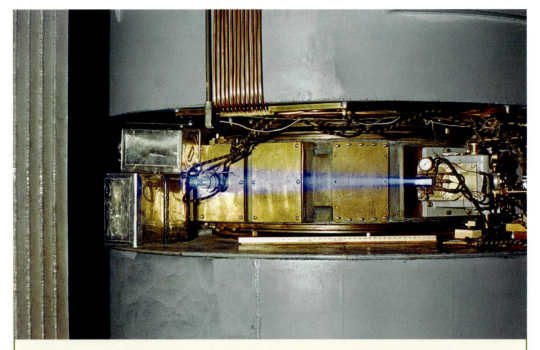

Ernest Lawrence's 60-inch (152-cm) cyclotron at the University of California, Berkeley. Lawrence was awarded a Nobel Prize for his work with cyclotrons and later developed a magnetic isotope separator used in the atomic bomb project. *(Lawrence Berkeley National Laboratory)*

machine, with a magnet weighing 220 tons (200 metric tons), and it would be used to discover carbon-14, neptunium, and plutonium. Lawrence's calutron was an industrial version of the magnetic mass spectrometer invented at the Cavendish Lab in Great Britain in 1918. It would be a critical component of the atomic bomb development project that would soon consume the United States.

When World War II started in Europe, the United States had all it needed to develop nuclear weapons and power systems. Available were many of the best scientists of Europe, an amazing array of homegrown physicists graduating from or teaching in universities, an industrial infrastructure capable of building anything that could be assembled by mankind, and a warehouse in New York filled with uranium ore. All that was needed was the spark to start the fire. In the next chapter, the spark and the beginning of the resulting conflagration are revealed.

6 The First Sustained Nuclear Power Production

In late 1939, with a world war already ignited in Europe, there was sufficient high-level theory to indicate that explosives using the binding energy of atomic nuclei were possible. Weapons based on nuclear energy principles had a frightening potential, with the most concern being that the opposing side would develop them first.

While solid theories made such devices seem possible, there were many details of implementation to be worked out. There were experiments to be run and data to be collected, and after that a massive industrial conversion would be necessary, going from small-scale laboratory setups to large-scale production. It was nothing that any individual, any organization, or any consortium of companies had the resources to make happen. There was too much risk of failure, even if the expense and effort were feasible. It was a problem of governmental scale, and not just any government could handle it. In this chapter, the process of awakening the U.S. government to engagement in this massive scientific research and development is examined.

The working weapon theory was that a runaway chain reaction of fissions could be started, but to this point no such action had been observed. It would take a large, pure sample of U-235 to perform an experiment, and the problem of purifying a rare isotope of uranium in mined ore was enormous. It would be a large waste of effort if it turned out that a chain reaction was not possible. This chapter reveals the important step of first proving the chain reaction concept using naturally occurring concentrations of the fissile uranium isotope, creating a controlled, nonexplosive form of nuclear energy release.

A LETTER TO THE PRESIDENT OF THE UNITED STATES FROM ALBERT EINSTEIN

By 1939, the United States had acquired a large group of expatriate European scientists, each associated with a university and engaged in research. All were unusually busy in the fast-breaking world of nuclear physics, exchanging papers, sharing experimental results, absorbing unsubstantiated rumors from Germany, and generally conspiring to somehow involve the government of their adopted country in large-scale nuclear research. It was not an easy quest, and there would be obstacles. In 1939, the U.S. military establishment was hardly the innovative powerhouse that it would later become, and a plea by a group of heavily accented theoreticians was unlikely to move mountains even if it promised a quick victory over potential enemies. The United States was not at war, and it intended to remain in that posture as long as possible.

Still, it was an effort that had to be made, as the European group became seriously concerned that the German Third Reich would develop nuclear weaponry. Leó Szilárd was particularly shrill in his activities, and at Columbia University he was doing what he could with limited research resources to find sustainable fission in uranium. Being naturally persuasive, Szilárd was able to borrow 500 pounds of black, dirty uranium oxide from the Eldorado Radium Corporation at Great Bear Lake in the Northwest Territories of Canada. Experiments showed him the value of crystalline carbon as a neutron *moderator,* and he managed to persuade the National Carbon Company of New York to make him some high-purity synthetic graphite. This specially made material was uncontaminated by the traces of boron that made most industrial graphite unusable for nuclear work. Even Szilárd's impressive skills at scrounging materials were far short of what was necessary, but he used his talents to persuade Enrico Fermi to stop in at the Navy Department on Constitution Avenue in Washington, D.C., on March 17, 1939, and talk to somebody about the urgent need and the great potential of nuclear power. The undersecretary of the navy was unavailable, but an appointment was made to see the technical assistant to the chief of naval operations, Admiral Stanford C. Hooper (1884–1955), the "Father of Naval Radio," and convince him of the importance of nuclear fission. This would be the first contact between scientists pursuing nuclear fission and the U.S. government.

The meeting did not go well. Although it was attended by an impressive audience of naval officers, men from the Bureau of Ordnance, and two scientists from the Naval Research Laboratory, Fermi's hourlong lecture on nuclear physics went over their heads. They seemed interested in a power source that required no oxygen, as they were thinking of submarine propulsion, but Fermi was too vague and unwilling to speculate. The navy was receptive to being included in announcements of success but not in participation. On July 10, Szilárd got the formal letter, thanking the scientists for the interesting presentation but denying any funding.

Helping and encouraging Szilárd in his sales efforts were two fellow Hungarians, Eugene Wigner and Edward Teller. Wigner was born in Budapest, Hungary, and in 1921 he became best friends with Leó Szilárd at the Technische Hochschule in Berlin, Germany. In the 1920s, he became deeply involved in quantum mechanics research and in 1929 was recruited by Princeton University, in New Jersey. Fellow scientists called him the "Silent Genius," and his work in nuclear reactor theory would be foundational.

Edward Teller was also born in Budapest, and as a child of 11 years he developed a powerful aversion to both fascist and communist governments. As a Hungarian nuclear physicist working in Germany in 1933, it was almost inevitable that he would wind up in the United States working on the atomic bomb program of World War II. He would be of immense value to the further development of nuclear weapons following the war. His strong political views and an insistence on making weapons large enough to evaporate an entire Pacific island would make him perhaps the most controversial scientist in the country.

Fermi gave up early and went back to his laboratory research, but the Hungarian team thought it was worth trying again. Thinking that they should try to approach this problem from an entirely different angle, they decided to draft a letter to Queen Elizabeth of Belgium. The rumors were that the Germans were buying uranium ore stockpiles from the rich lode of pitchblende in the Belgian Congo in equatorial Africa. The right word to the queen might persuade her to discourage her countrymen from uranium commerce with the Third Reich, and stopping German research was as important as starting research in the United States. Unfortunately, none of them knew the queen of Belgium, but they knew someone who did. Albert Einstein had met her in 1929, and they were in regular communication. Einstein was a world-famous scientist, who had left his native Germany for a position at Princeton's Institute for Advanced Study. It was his lofty theories of mass and energy that gave credence to this hypothetical nuclear weapon.

Wigner and Szilárd made an appointment to see Einstein and then drove to his summer home on Nassau Point in Long Island, New York, on Sunday, July 16. Szilárd had never learned to drive, so Wigner controlled the car as they strategized and plotted all the way there, carrying a draft of the letter. Einstein was all for the letter, but he hesitated to burden the queen of Belgium with the problem of sales with Germany. He counterproposed a note to the Belgian ambassador, and Wigner, having an inkling of knowledge of diplomatic protocol, suggested a cover letter to the State Department. They worked all day on several drafts. When Szilárd got home, he found a message from a Dr. Alexander Sachs.

Alexander Sachs (1893–1973) of the Lehman Corporation was an economist, a banker, and a personal friend of the president of the United States, Franklin D. Roosevelt. He had learned of Szilárd's letter mission from a mutual friend, and he had a suggestion for improvement. The letter should be rewritten as a mandate for the United States, and it must be delivered directly to the president. Szilárd eagerly went to work on a new draft, and he made another appointment with Einstein.

Wigner had gone to California for a vacation, but Szilárd recruited Teller to drive to Long Island in his trusty 1935 Plymouth on Sunday, July 30. After careful rewriting and editing, the three scientists had crafted a longer, two-page version. It begins with the following:

> Sir:
> Some recent work by E. Fermi and L. Szilard, which has been communicated to me in a manuscript, leads me to expect that the element uranium may be turned into a new and important source of energy in the immediate future. Certain aspects of the situation which has arisen seem to call for watchfulness and, if necessary, quick action on the part of the Administration. . . .

The letter continues, advising that a nuclear chain reaction can possibly be used to generate vast amounts of energy and for extremely powerful bombs. Further suggested is governmental funding of research, under an appointed administrator, coordinating university-based work with industrial laboratories, and the letter ends with a hint that Germany may

THE NEED FOR SECRECY

With the U.S. formal engagement in World War II and the elevation of nuclear research to highest priority, secrecy of all aspects of the project became absolute. First, no physicist could publish papers having to do with nuclear fission, so as not to reveal to external governments that bomb research was proceeding. The sudden stoppage of any nuclear publications only indicated to the Soviet Union, always sensitive to Western secrets, that something was going on in the United States, and they immediately started an infiltration effort.

Immense laboratory complexes were built in remote, uninhabited locations and given names "Site X" and "Site Y." Workers never knew what they were building. Industrial workers at production facilities were never told what they were producing and were forbidden from talking to the workers standing next to them, much less anyone outside the building or from another building. There were words that could not be said. No one could say uranium-235. The material was referred to as "oralloy." One could never refer to the uranium isotope separation plant as anything but Y-12, K-25, or S-50. Plants and laboratories were ringed with high fences, barbed wire, and armed watchtowers. Photography was highly restricted, and every piece of paper had to go in a combination lock safe or vault if it was not being written on or read from. There was a counter-story written to explain every odd sound, flash of light, or truck leaving a hidden facility. Foreign scientists working on the project, of which there were many, could find themselves being tailed by security operatives if they left the laboratory grounds.

Scientists were given working names, which were printed on their security badges. Enrico Fermi was Mr. Farmer and Niels Bohr was Mr. Baker. Once Bohr wound up at a checkpoint without his badge, but Fermi was able to vouch for him, saying, "I assure you that this is Mr. Baker, sure as my name is Mr. Farmer." Scientists disappeared from their university positions and were not heard from again until after the war. The general public in the United States had no idea that atomic weapons were being developed in their country. Knowledge was dispensed only on a strict need-to-know basis. The vice president of the United States, Harry S. Truman (1884–1972), was not advised that his country was developing an atomic bomb until he was sworn in as the commander in chief on Franklin Roosevelt's death on April 12, 1945. Neither the Germans nor the Japanese knew anything about the U.S. bomb project and did not know to be concerned.

(continues)

(continued) _____

The British were also very aware of the need for atomic security, and they carefully monitored all correspondence and radio traffic into and out of the home islands. When Copenhagen was occupied by German forces, Niels Bohr sent a message to a German physicist working in England, mentioning a "Miss Maud Rey at Kent." Thinking that it was surely a coded message, the British code masters found that the letters could be rearranged to form "radium taken." Miss Maud Rey turned out to be the former governess of Bohr's children, and she was indeed living in Kent. There was no coded message. The MAUD Committee was named in her honor.

already be engaging in nuclear work. On August 15, Szilárd transmitted the document to Sachs with the authoritative Einstein signature.

Sachs knew that the concept had to be sprung on the president at just the right time, when he was in the correct mood, and unfortunately events in Europe were monopolizing his attention. Weeks passed. Finally, as Szilárd's patience was at the boil, Sachs saw the opportunity to approach the president and present the letter on October 11 at the White House in the Oval Office. Neither man actually read the letter, but Sachs gave Roosevelt an 800-word synopsis of what he had learned from Szilárd.

Roosevelt absorbed it all and summed it up, saying, "Alex, what you are after is to see that the Nazis don't blow us up."

"Precisely," said Sachs.

Roosevelt called in his aide, saying, "This requires action."

So began the largest scientific research and development program in history. It would take another letter from Einstein five months later on April 25, 1940, to speed things along, but the ball was now rolling, and there was no stopping it.

THE FIRST NUCLEAR REACTOR

An unused football stadium, Stagg Field, with abandoned squash courts underneath the west stands at the University of Chicago was the perfect setting for a large, secret experiment. It was not a place anyone would look for world-changing science. Although preliminary studies of a graphite-moderated fission reaction had been studied at Columbia and Princeton, the groups were combined and sent to Chicago for the definitive experiment by the sponsoring agency, the Office of Scientific Research and Development, directed by Arthur H. Compton.

With his methodical, thorough experimental ethic, Dr. Enrico Fermi directed the project, building 30 test assemblies out of larger and larger piles of pure graphite bricks, with interspersed cylinders of pressed uranium oxide. Everything was carefully considered and designed, from the allowable impurities in the graphite to the radius and length of each uranium cylinder, to the optimum spacing between uranium pieces, with calculations improving as larger and larger subcritical piles were assembled, observed, and dismantled. On November 16, 1942, when the team thought they had enough data to predict the size of a fully critical, continuously running nuclear reactor, they started stacking layers of graphite and uranium on the wooden floor of the squash court, building "CP-1," or Chicago Pile number 1.

Chemically pure graphite seemed the ideal moderator material to slow the neutrons down to fission speed. Under severe badgering from the eccentric Hungarian genius, Leó Szilárd, three companies were encouraged to produce some remarkably pure synthetic graphite. On the German side of the race to produce nuclear power, graphite had been quickly dismissed as a possibility. In Europe, graphite was mined, for use in producing pencils, and graphite was readily available, but its mineral impurities made it unusable. The Germans chose to use the exceedingly rare material deuterium oxide, or heavy water, as a moderator.

The Chicago pile was big. It was roughly a sphere, 25 feet (7.6 m) in diameter, looking like a huge basketball made of black Legos. Wooden timbers held up the bottom half of the thing, and the cadmium control-rods, intended to absorb neutrons to selectively kill the fission reaction, ran in from the front, through channels routed in the graphite. Electronic radiation counters and controllers for the *control rod* motors were piled up in the squash court balcony. On December 2, 1942, the reactor was fully built, checked out, and ready to be tested. Forty-two men and one woman, Leona Woods (1919–86), crowded onto the squash court balcony and a couple of stations on the pile to see the first sustained nuclear reaction. George Weil stood on the floor in front of the pile to move one critical control rod by hand. Harold Lichtenberger, W. E. Nyler, and A. C. Graves got to stand on a wooden platform in back of the reactor, holding carboys full of cadmium salt solution, poised to drop them on the graphite and kill the reaction in case things got seriously out of control. Fermi,

Process of Chain-reacting Nuclear Fission

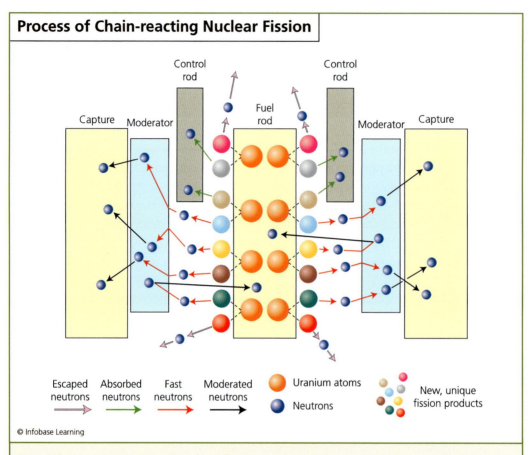

Escaped neutrons →
Absorbed neutrons →
Fast neutrons →
Moderated neutrons →

Uranium atoms

Neutrons

New, unique fission products

© Infobase Learning

In a nuclear reactor, neutrons born from fission are slowed down to a crawl by traveling through a moderator, such as graphite. They can then cause other fissions or be lost to a control rod that is meant to absorb neutrons, or they can simply leak out of the reactor.

Compton, Herb Anderson, and Walter Zinn, from Canada, sat at the control desk. Norm Hilberry stood ready with an axe to cut the rope holding the zip rod, which would fall by gravity into the pile if they needed a quick shutdown. Everybody else was there to watch.

For most historical scientific experiments and discoveries, we have only a sketchy, incomplete account of the detailed activity, and we must rely on recollections and cleaned-up summaries of what happened. The world's first reactor startup is a very rare exception. Neither recordings nor photographs were made, due to the military secrecy of the operation, but Leona Woods, the youngest person to witness the event, took detailed, minute-by-minute notes. Fermi was not a very talkative person, and the gallery remained mostly in awed silence. Woods's written record of each action and spoken word is considered correct.

At 9:45 A.M., the experiment began. Fermi called for withdrawal of the electrically driven control rods, somebody threw the switch, and the crowd hushed as the DC motor whined. Out slithered the cadmium rods from the middle of the slippery graphite pile. All eyes turned to the neutron counter dial and the pen chart recorder, which was keeping a record of the neutron activity in the pile on a continuous roll of paper, with a motor-controlled blue pen. The count rate stepped up a little, but nothing to write a paper about. An audio amplifier and a speaker were connected

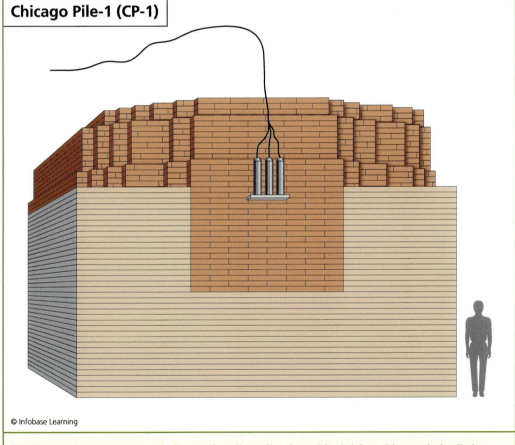

Chicago Pile-1 (CP-1)

© Infobase Learning

The first nuclear reactor ever built was literally a pile of graphite bricks, with rounded cylinders of uranium evenly dispersed throughout. The three cylinders wired in front are neutron detectors used to monitor the fission activity.

to the counting equipment, giving an occasional click sound as a neutron strayed into the detector tube.

Fermi, a man of few extraneous words, said, "Zip out." It was just after 10:00 A.M. Zinn pulled out the gravity rod and tied its rope to the balcony. The neutron count rate rose noticeably.

It was 10:37 A.M. With his eyes locked on the neutron rate dial, Fermi called to Weil, "Pull it to 13 feet, George." Weil, standing at the reactor face, carefully withdrew his small vernier control to 13 feet (4 m), marked on the side of the rod. The neutron count jumped. The crowd murmured, as slide rules slid and pencils scratched.

"This is not it," predicted Fermi. "The trace will go to this point and level off." He pointed to a blank spot on the pen chart. Slowly the pen moved up and leveled off, right where he said it would. The crowd was enraptured and studied the new flatline for seven minutes, then Fermi called to Weil for another foot of rod. Weil complied. The count rate increased but leveled out. For nuclear fission to be happening at a useful level, the rate would have to increase exponentially, tending to become a vertical line on the graph.

At 11:00 A.M., Fermi, seeming not in the slightest way impatient, called to Weil for another six inches of vernier control. At 11:15 A.M., a little more. At 11:25 A.M., another smidgen. After each movement, the count rate would increase slightly, and the clicks of neutrons hitting the detector tube, amplified and put over a loudspeaker, became irritating. Fermi seemed to be enjoying the drama, as he correctly predicted each level out of the pen on the chart. He knew they were getting close. Just to be absolutely sure of things, Fermi ordered that the automatically actuated control rod be dropped in, to test the circuit. The safety rod banged home, and the count rate dropped abruptly, just as it should.

Satisfied, Fermi called for a restart, and at 11:35 A.M. the safety rod was reset in out position, and the vernier was carefully pulled out a little more. The count rate rose and rose. The crowd watched and waited, silent, enthralled by the rising neutron count. Suddenly, there was a loud bang.

Everybody froze. Then, as nothing seemed to be melting through the floor, they realized that the safety rod had automatically tripped, sending it rapidly into the reactor core to stop the reactions. The tripping point was set too low on the neutron rate meter. It was quickly adjusted.

Fermi announced that "I'm hungry. Let's go to lunch." Weil parked the vernier rod, the motor rods were driven in, the zip was lowered in, and the party broke for the dining hall.

Over lunch, not a word was said about neutrons, graphite, or the unspeakable substance uranium. Fermi just ate lunch, giving not a hint of a pep talk, as the others went on about anything except the "game."

At 2:00 P.M., they assumed their positions in the squash court, and it took 20 minutes to warm up the equipment and withdraw the safeties to their previous condition. "All right, George," called Fermi. Weil took this to mean restore the vernier to its last position, and he did so. The count rate was high.

At 2:50 P.M., Fermi called for another foot of rod. Out it came. The pen chart ran off the top of the graph, but they still weren't exponential. Somebody clicked the pen chart range up by a factor of 10, to get the pen back on the chart. Everybody watched.

At 3:20 P.M., Fermi said "Move it six inches." Weil pulled six inches. The random ticking sound on the speaker, indicating individual neutrons counted, was becoming frantic. What was once a series of clicks now sounded like air escaping or waves crashing on rocks. Over the hiss from the speaker Fermi called "Pull it out another foot." Weil pulled.

Fermi, supremely confident, turned to Compton and said, "This is going to do it. Now it will become self-sustaining. The trace will climb and

The Birth of the Atomic Age by artist Gary Sheahan (1893–1978). Although the room in which the first nuclear reactor was built was too crowded to take a picture in, this painting captures the moment of the first self-sustaining chain reaction at the University of Chicago in 1942 *(Chicago History Museum)*

continue to climb. It will not level off." He pulled his slide rule and began calculating. He flipped his rule over and penciled temporary numbers on the back. He looked grim, as the count rate rose.

Three minutes later, Fermi made another calculation, and the crowd was jostling for position to see the count rate on the chart. Wilcox Overbeck began calling out the numbers from the chart as the pen traversed. Fermi, stone-faced and a picture of calm throughout the exercise, suddenly closed the C-scale on his slide-rule and grinned broadly. "The reaction is self-sustaining," he announced. "The curve is exponential." The time was 3:52 P.M. on December 2, 1942. Power production by the direct conversion of matter to energy had been proven feasible.

It was a closely held secret. The construction of CP-1 was finally declassified on May 18, 1955, when Enrico Fermi and Leó Szilárd were awarded the patent, number 2,708,656, for the nuclear reactor. The people of the United States were given their first glimpse of what their tax money had paid for in 1942. The experiment cost about $1 million, or, adjusted for inflation, $12.5 million.

THE *MANHATTAN PROJECT* BEGINS

On September 13, 1942, an important meeting of the Project S-1 Executive Committee was held. The United States was now fully engaged in the war, and it was time to move the atomic bomb project from a cautious and tentative rate of progress to full speed forward. Present at the meeting were Arthur Compton (1892–1962), director of S-1; Lyman Briggs (1874–1963), the former chairman of the Uranium Committee; James Conant (1893–1978), president of Harvard University; Ernest Lawrence, the cyclotron expert from Berkeley; Eger Murphree (1898–1962), petroleum chemist at Standard Oil; and Harold Urey (1893–1981), another chemist, but with a Nobel Prize. Complete secrecy was necessary, so the meeting was held deep in a forest in Monte Rio, California, in front of the massive stone fireplace in the clubhouse of the Bohemian Grove.

The Grove, populated by undisturbed redwoods more than 1,500 years old, is owned by the Bohemian Club, an exclusive, extremely secretive men's club founded in 1872. Over the entrance is carved the motto: "Weaving Spiders Come Not Here." A more secure venue could not be found. The urgent topic of discussion was fast neutrons.

A report sent from the MAUD Committee, the top secret British study of nuclear fission, had just come to light, and it detailed an important finding. It had been known since 1939 that uranium-235 would fission under the influence of neutrons slowed to thermal speed, but there was another point on the spectrum of neutron energies where fission would also be initiated. It was at the opposite end of the range, where 1 MeV neutrons, fresh from the fissioning event, would also cause energy-releasing fissions. This fact was key to the development of nuclear weapons, because it meant that a bomb could be built small and light and carried by an airplane. Until then, it had been thought that any chain-reacting uranium explosion would have to be graphite-moderated, and the weapon would be the size of a small house and weigh many tons. Without the moderator, and using pure uranium-235, the explosive core could be as small as a pineapple.

The men in the meeting understood the implications of this new information, but they also realized that they were in over their heads. They decided to establish a new, centralized laboratory to do nothing but study fast neutrons. For the next couple of days, it was code-named Project Y, and they decided it should be run with military precision, speed, and efficiency by a West Point man. Four days later, Colonel Leslie Richard Groves (1896–1970) of the U.S. Army Corps of Engineers was assigned to run this new project. Born in Albany, New York, Leslie Groves was educated at the University of Washington, the Massachusetts Institute of Technology, and West Point, where he was fourth in the class of 1918. He had just completed a huge construction project, building the Pentagon military office building in Virginia, and was looking forward to a vacation.

Instead, he was promoted to brigadier general and handed a project called "Laboratory for the Development of Substitute Materials," a name chosen for its misdirecting properties. Groves did not like the name. He changed it to the "Manhattan Engineers District," for a nonexistent office of the Corps of Engineers, and seven days later he bought 52,000 acres (210 km²) of land in rural Tennessee, hidden between mountain ranges, called Oak Ridge. It would be given the prosaic name "Site X," and his project would be headquartered there, far from prying eyes. Everyone except Groves would call it the "Manhattan Project."

Nuclear physics was now on the fast track. Decisions and directives that had taken months to become effective before the war now took minutes to be implemented, and budgets that were in the thousands of dollars were now in the billions. New cities would be built from scratch in weeks, universities would be drained of science and engineering faculties, and even silver would be shipped out of the U.S. Treasury by the ton. It would be an industrial application of pure science as the world had never seen, and it would be carried out in secret. Unlike Germany or any other place in Europe, the manufacturing plants and laboratories would not be subject to bombing, sabotage, or enemy observation.

The next chapter covers the intense three-year effort, beginning with construction projects in Tennessee, Washington State, and New Mexico, and ending with the unconditional surrender of the Empire of Japan. This unusual, government-sponsored endeavor would forever skew nuclear power, reminding everyone of bombing and unleashed radiation, but it would also give it a tremendous push.

7 Nuclear Weaponry Development

Brigadier General Leslie R. Groves was presented with an enormous task. He had to lead and coordinate a project that would take newly discovered principles and theories of nuclear interactions and translate them into manufactured military weapons. The work had to be done under emergency conditions and in total secrecy, using scientists recently imported from enemy-held territory. He had multiple tasks to be done first, before anything else was done. He had to build three city-sized laboratories: Site W, in the desert in Washington State at Hanford on the Columbia River, would be built to produce usable quantities of an element that had never before existed, named plutonium. Site X, in Oak Ridge, Tennessee, would be built to separate a rare isotope from tons of uranium ore, one atom at a time. Site Y, built in the high desert in New Mexico at Los Alamos, would be the intersection point, where these two exotic materials would be fashioned into a new type of extremely powerful bomb.

Groves needed experts in fields that were not even invented yet. He needed people, materials, and money at a time when all three were in short supply, but most immediately he needed help. He saw a requirement for a top theoretical scientist in a management position to keep the continued nuclear research and development running fast and efficiently. His Manhattan Project needed a scientific director, and he chose Dr. J. Robert Oppenheimer (1904–67), a physicist from the University of California, Berkeley.

Oppenheimer and Groves were complete opposites. Groves was large and overweight and fond of eating a certain type of candy called turtles. He was a career military man, trained in the United States as an engineer, with experience in large construction projects. His political views were conservative, and his manner could be brusque and direct. Oppenheimer was thin and underfed, and eating was not his favorite activity. He was a career academic, trained at the University of Göttingen, and his experience was theoretical physics. His political views were socialist with communist leanings, and his manner could be arrogant and sarcastic. They were a perfect match. Both became driven by the project. The two became good friends. When the war closed, they would share the credit for having led a magnificent job and the blame for having unleashed a dangerous product.

FIRST WORK AT THE LOS ALAMOS LABORATORY

The program to develop nuclear weapons was designed with two parallel paths, and only one had to succeed. Path 1 was to build a weapon using uranium purified to at least 80 percent fissile U-235. In path 1 were two parallel, independent sub-paths, designated Y-12 and K-25, to be implemented at Site X, or the Clinton Engineering Works at Oak Ridge. Y-12 was to separate U-235 from natural uranium using industrial-grade magnetic mass spectrometers, or "calutrons," developed by Ernest Lawrence at the University of California, Berkeley. In a calutron, uranium atoms are ionized in a vacuum chamber and accelerated into a negatively charged electrode. In the flight path is a powerful magnetetic field, turning the flying ions through an angle of 180 degrees. The heavy U-238 ions have trouble making the turn, and the lighter U-235 ions are more likely to hit the electrode and be collected. The process was extremely slow, and managed to make enough U-235 concentrate in two years running 24 hours a day, seven days a week, to make exactly one bomb.

The priority of the bomb project is illustrated by Groves's ability to commandeer materials. To build the alpha and beta calutrons, enormous magnet coils were needed. Sufficient copper of adequate purity was simply not available during the war. Copper supplies were so short that pennies were being minted from steel. Groves asked his scientists what material could be substituted for pure copper. They told him silver would do nicely. Groves arranged to have 14,700 tons (13,300 metric tons) of pure silver transferred to the Manhattan Project from the U.S. Treasury.

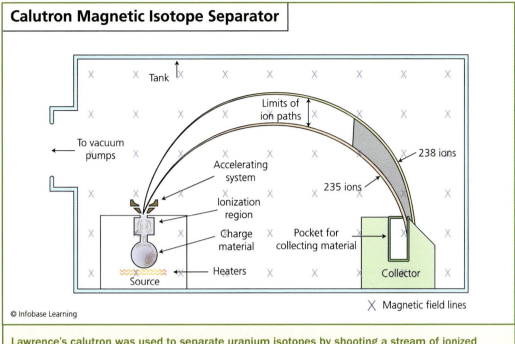

Calutron Magnetic Isotope Separator

Lawrence's calutron was used to separate uranium isotopes by shooting a stream of ionized uranium particles through a magnetic field. The stream was bent in the field, but the lighter uranium-235 ions were bent at a shorter radius than the heavier uranium-238 ions.

The second sub-path for U-235 concentration was K-25, which was an entirely different enrichment scheme. K-25 used gaseous diffusion, in which uranium hexafluoride gas is pressurized against a porous membrane. Gas molecules containing U-235 travel slightly faster than molecules containing U-238 and are more likely to diffuse and make it through to the other side of the membrane. By passing the gaseous uranium through hundreds of stages of diffusion, large concentrations of U-235 can be achieved. The K-25 plant was the largest building on earth. It covered more than 2,000,000 square feet (609,600 square meters) of ground, cost $512 million dollars to build in 1944, and employed 12,000 full-time workers. For all its great potential as a uranium enrichment facility, the K-25 plant did not contribute any *enriched uranium* to the atomic bomb. By the time it was operating and producing enriched product, the war was over, but it continued operations until 1987 and made a great deal of enriched uranium for the nuclear power industry. After the end of the war, with the K-25 plant running at full speed, the Y-12 plant was shut down and dismantled. The silver was given back to the Treasury Department, with expressed gratitude.

Path 2 of the Manhattan Project was to build a nuclear bomb using *Pu-239*, plutonium, as the fissile material. Using plutonium had one enormous advantage: It did not require *enrichment*. The disadvantage was that it could not be mined, as it was not known to exist in nature. It had to be produced, and the only way to produce it was by bombarding natural uranium with neutrons. The U-238 in natural uranium would activate into U-239 upon neutron capture, and this would beta decay into neptunium-239, which was unstable and would quickly beta decay into relatively stable Pu-239. In beta decay, a neutron transforms into a proton, and the decaying isotope retains its mass number while changing its atomic species.

Plutonium had been made at the Berkeley cyclotron in vanishingly small quantities, but for use in a bomb several kilograms would be necessary. The way to bombard U-238 with large neutron flux was to use a nuclear reactor, running at very high power. Any neutrons produced in the fission action that were not used to produce continuous fissions would have a good probability of being captured by U-238 atoms in the fuel, converting the otherwise useless U-238 to *fissile* Pu-239.

Enrico Fermi led a team at the new Argonne Laboratory to quickly design a reactor for the Hanford Works that could produce 250 megawatts of power. Each person in the United States uses an average of 1,400 watts of power, 24 hours a day, so the power output of this plutonium production reactor, named *B-Reactor,* could have produced enough power for 180,000 people. Power production, however, was not the goal of this reactor, and all the energy was exhausted into the Columbia River on the flat desert in central Washington State. The goal was simply to make neutrons and use them to convert U-238 into Pu-239.

B-Reactor for Site W was an upscale version of CP-1, still using the natural uranium fuel and the highly efficient moderator made of pure graphite pressed into bricks. The uranium was formed into slugs, each about the size of a roll of quarters, sealed in aluminum casings. The fuel slugs were lined up in 1,500 aluminum tubes, running horizontally through a 1,200-ton (1,089-mt) graphite cylinder, 28 by 36 feet (8.5 by 11 m), lying on its side. River water was pumped through the aluminum tubing at 30,000 gallons (110,000 L) per minute to remove the energy from the reactor. The industrial complex supporting the plutonium production at Hanford, Washington, was about half the size of the state of Rhode Island, with 512 new buildings as well as temporary housing. After two years of construction, involving 42,400 workers, on Tuesday evening, September 26, 1944, B-Reactor was ready to be powered up and start producing plutonium.

E. I. du Pont de Nemours and Company was responsible for the operation and engineering of this enormous chemical plant. Enrico Fermi and a crowd of du Pont executives were on hand at the Hanford Works to watch the fuel being loaded and to observe the start-up of the new reactor. Everything checked out perfectly, and the operations staff withdrew the control rods in stages, just as Fermi had done in the first reactor experiment in Chicago. The pile went critical, with a self-sustaining chain reaction, at about midnight. Cautiously the power was increased to operating level, and by 2:00 A.M. the reactor was operating smoothly at high power, with cooling water flowing through the aluminum tubing.

All went well for about an hour, and then the operation took an unanticipated turn. For some reason, the operators had to begin stepping out control rods to maintain the critical condition, in which the number of neutrons lost to leakage, capture in U-238, or fission exactly equaled the number of neutrons produced in fission. Eventually, the operators ran out of control rods to pull, and B-Reactor died early Wednesday morning.

Fermi tried to remain calm and think through the failure, considering all possible reasons why the reactor had quit generating power. Groves, needless to say, was not happy with the unexpected behavior of B-Reactor and neither was du Pont. A theoretician from Princeton University, John A. Wheeler (1911–2008), stepped forward with an explanation. Using the time required for the effect to shut down the reactor as a clue, Wheeler predicted that iodine-135 was being produced as a fission product, with a radioactive half-life of 6.68 hours. There was nothing wrong with having iodine-135 in the reactor, but it decays into xenon-135, which is a voracious neutron absorber, 150 times more effective at neutron capture than the cadmium metal used in the control rods. The Xe-135 would decay with a half-life of 9.13 hours into something harmless, but meanwhile it was being made by the decay of I-135, and it was shutting the reactor down. Fermi and his team had not observed "xenon poisoning" in their low-power experiments because only at high power was enough I-135 produced to lead indirectly to shutdown.

The solution to this problem of high-power shutdown was to build into the reactor enough excess ability to fission, or reactivity, to overcome this poisoning effect. Fortunately, Wheeler had insisted that du Pont build more fuel channels into the graphite pile than would be necessary to sustain a reaction, adding millions of dollars to the cost of the project, just in case something unforeseen came about. The extra tubes would accommodate another 504 lines of fuel slugs, and this was sufficient excess reactivity to overcome the xenon effect. Two additional graphite piles, the D-Reactor and the F-Reactor, were started up in December 1944 and February 1945, using the modified fuel loading of 2,004 tubes. By April 1945, plutonium was being chemically separated from spent fuel slugs from the three production reactors and shipped to Los Alamos every five days.

Nuclear physics lived at Site Y, or the Los Alamos Laboratory, on a mesa in northern New Mexico. The site was chosen by Oppenheimer and Groves as an out-of-the-way location, far from prying eyes and with plenty of room to experiment with explosives and radioactive materials. It was the location of a private boy's academy named the Los Alamos Ranch School. On November 22, 1942, the Corps of Engineers submitted an appraisal of the site, which was found perfect, and the U.S. government bought it for $440,000. Construction of a large laboratory and living complex began immediately, and Oppenheimer crisscrossed the country recruiting nuclear scientists from every major university. Enrico Fermi, Hans Bethe, Edward Teller, Stanislaw Ulam, Seth Neddermeyer, George Kristiakowsky, and a host of other extremely capable physicists agreed to move to Los Alamos for the duration of the war. Leó Szilárd, who was not acclimatized to remote, rustic conditions, balked. "Nobody could think straight in a place like that," he complained. "Everybody who goes there will go crazy." Isador I. Rabi (1898–1988), a very capable theorist from Hungary, also declined the invitation, thinking that he was more useful at the radar laboratory at MIT. "I'm very serious about this war," he said. "We could lose it with insufficient radar."

Oppenheimer needed scientists, support personnel, and equipment for neutron research. He managed to wrangle the cyclotron from Harvard University and two Van de Graaff linear particle accelerators from the University of Wisconsin to be used for artificially generating neutrons. In April 1943, the laboratory opened, with a school set up to indoctrinate the incoming scientists and reveal to them the purpose of the new facility. The lecturer was Robert Serber (1909–97), and the subject of his five-day course was the production of a practical military weapon. In the secret parlance of the time, it was referred to as the "gadget."

The gadget was to bring together two subcritical pieces of fissile material quickly to form a hypercritical assembly. A hypercritical block of material would contain enough fissile U-235 or Pu-239 to form more than an assembly that was critical, in which the neutron production and loss are balanced. It would be a runaway chain reaction in enough fissile isotope to be critical three or four times, and the uncontrolled nuclear fission reaction would proceed to increase at an explosive rate.

A nuclear power reactor and an atomic bomb both use the chain reaction of neutrons causing fission, but the goals and the methods are different. In a power reactor, neutrons are slowed down to thermal speed, or the speed of normal molecules and atoms, and are captured by uranium or plutonium nuclei, causing fission. It takes time to slow the neutrons down, and all the neutrons thrown out in the fission process, about 240 per 100 fissions, are not thrown out of the fission debris at once. Some take several seconds to pop out, and these factors slow down the response of a nuclear reactor to changes in *criticality*. This is an attractive feature, because it renders a nuclear reactor easy to control. The response time at the controls is slow and smooth.

There are therefore two modes of criticality. There is delayed criticality, in which all the neutrons, including the ones delayed from fission, are used to balance with the lost neutrons. There is also prompt criticality, in which only the instantly available neutrons are counted, and this requires a larger mass of fissile material, as there are fewer neutrons available promptly to contribute to fission. Prompt fission is essentially instantaneous, with no built-in delay to moderate the controls.

There are also two modes of fission. Thermal fission is used in power reactors, as it makes optimum use of available neutrons and it involves a slowing-down medium, or moderator, inserted between fissile parts, and this moderator can be used conveniently as a coolant. Fast fission uses only high-speed neurons, running at about 1 MeV, as they are thrown out in the fission process and before they hit anything to slow them down.

Power reactors use delayed, thermal fission. Atomic bombs use prompt, fast fission. A power reactor makes a slow climb from full shutdown mode to a critical, self-sustaining power level, and any change in power level is made slowly. It took two hours for B-Reactor to move from zero to full operating power, which gave the auxiliary equipment such as coolant pumps plenty of time to adjust to the high-power conditions. An atomic bomb, or gadget, goes from zero power to a massive power spike in less than a millisecond, or one one-thousandth of a second. The criticality is promptly achieved on the high end of the neutron energy spectrum so that no time is wasted slowing down or diffusing through intervening moderator, and the reaction is completely out of control, with no effort to balance the losses and productions of neutrons. More neutrons are produced and wasted than could ever be used for fission, and the mechanism is destroyed in a momentary burst of extreme energy.

TWO ATOMIC BOMB DESIGNS DIVERGE

Various designs for this proposed weapon were debated, experiments were performed, and nuclear physics was given a thorough workout. The bomb had to be small and light enough to be delivered to the target by a large airplane, and it would be deployed by gravity, by dropping it from a high altitude. Simplicity was a goal, and eventually a design that would work using either uranium or plutonium emerged.

It was code-named "Thin Man." It was 17 feet (5 m) long, with a bulge on the end. It looked like a telephone pole with fins on the end. The bulge at the end housed a subcritical cylinder of fissile material, with a three-inch (8-cm) hole in the center. A second subcritical piece of fissile material, cylindrical and just big enough to fit in the hole, was kept at the other end of the bomb, in a gun barrel running from end to end. To set off the bomb, an explosive charge behind the subcritical projectile would accelerate it to 3,000 feet per second (914 mps). As it passed through the subcritical component in the bulge there would suddenly be enough material in one place to make a hypercritical assembly of fissile uranium or plutonium, and the resulting reaction would run away explosively.

By May 1943, plutonium was arriving at Los Alamos, and an unforeseen problem with using plutonium in a bomb became clear. The plutonium-239 supplied from the production reactors had a slight contamination of plutonium-240. The *Pu-240* had a tendency to fission spontaneously, and the gun-barrel design could not send the two subcritical pieces of plutonium together fast enough. The spontaneous fissions would set off the assembly as the projectile proceeded down the gun barrel, and the bomb would come apart before it was able to achieve hypercriticality. It was impossible to chemically separate Pu-239 from Pu-240, and an isotope separation as was being used to make U-235 was out of the question. The masses of the two plutonium isotopes were too close together.

There was danger of the entire plutonium production effort becoming useless, but there was another way to assemble the plutonium into a hypercritical mass. An American physicist from Caltech, Seth Neddermeyer (1907–88), had been pushing an idea called *implosion,* and he presented his first technical analysis of the idea in April 1943, just as the

ESPIONAGE IN THE LABORATORY

Secrecy in the Manhattan Project was tight and very well managed. U.S. enemies in Japan and Germany had no idea what was going on in universities, government laboratories, and industrial plants spread all over the country. The Japanese government did not know we had an atomic bomb until we dropped one on them, and even then there was skepticism. The German scientists were told after the war that we had developed the bomb. Still thinking in terms of a large nuclear reactor, they found it hard to believe that we had an aircraft big enough to carry such a bomb. Even most people who worked on the bomb project were surprised when the atomic bomb development was announced. Construction crews, lab technicians, and industrial workers were all kept in the dark. Laura Fermi, the wife of Enrico Fermi, did not know what her husband had been working on for four years until he gave her a copy of the book *Atomic Energy for Military Purposes* by Henry DeWolf Smyth in September 1945.

Nobody knew what the United States was working on, with the exception of the Soviet Union. The Soviets, apparent masters of spy craft, had infiltrated the Manhattan Project at multiple points and gained enough information to quickly duplicate the methods and designs for which the United States had worked so hard, and they developed similar weapons with minimum effort. We did not know exactly how much information the Soviets had extracted from the bomb program until after their government collapsed in 1990, and the files of their Committee for State Security, or KGB, were opened. The Soviet infiltration turned out to have been deeper and earlier than was realized. The information was sufficiently detailed for the Soviet scientists to build a duplicate copy of the CP-1 reactor in Chicago, but in translation their information was slightly garbled. The CP-1 was built in the abandoned squash courts under Stagg Field. The Soviet documents say that it was built in a "deserted pumpkin patch."

The primary information leak point may have been Klaus E. J. Fuchs (1911–88). Fuchs was born in Rüsselsheim, Germany, to a Lutheran pastor Emil Fuchs and Else Wagner. He attended both Leipzig University and Kiel University and joined the Communist Party of Germany in 1932. Finding himself at odds with the Nazi government, he fled in 1933 and was able to land in Bristol, England. Fuchs earned a Ph.D. in physics at the University of Bristol in 1937 and got a teaching job in Edinburgh, Scotland, the same year. Although he was interned at the beginning of World War II for being a

(continues)

(continued)

German citizen, the British needed every nuclear physicist they could find for a possible atomic bomb program, and he was granted British citizenship in 1942, signing the Official Secrets Act.

In late 1943, the small British bomb program allied with the Manhattan Project, and the British scientists were loaned to the United States. Fuchs was first assigned to Columbia University in New York City, but in August 1944 he was moved to Los Alamos Laboratory to work in the theoretical physics division. There, he had access to the difficult problems that were being solved for the bomb design, particularly the implosion method being developed for Fat Man, and he passed all of it to his Soviet contacts.

In 1946, after passing information concerning the secret development of the hydrogen bomb to the Soviets, Fuchs returned to Great Britain, where he was confronted by intelligence officers. An effort to crack Soviet ciphers, known as the Venora Project, had implicated him as a spy for the KGB. Finally confessing in 1950, Fuchs was tried and convicted of passing military secrets. His entire trial lasted 90 minutes.

plutonium crisis became evident. Everyone was familiar with the action of a chemical explosive. Set off a spherical bomb, and a shock wave radiates out from the point of explosion. If the spherical bomb is made hollow, with a void in the center, then two shock waves are produced. Still the outer shock wave radiates outward, becoming bigger and more diluted as it expands outward. The simultaneously generated shock wave at the center radiates inward, becoming smaller and smaller and more concentrated the farther it develops, until it is an extremely intense, spherical pressure wave at the center of the bomb.

Neddermeyer suggested that this inner shock wave could be used to shrink a small sphere of plutonium very quickly. The sphere would be so small as to be subcritical, not having sufficient material to form a *critical mass*. The size of a hypercritical mass depends on several factors, such as the shape of the mass, the number of atoms of fissile material present, and the distances between fissile atoms. The distance between two atoms in a block of plutonium would seem fixed. It is, after all, a solid, incompressible piece of metal. However, in the extreme forces produced by an implosion shock wave, metal can actually be compressed. For just an instant, a piece of solid metal the size of a softball can be compressed to something the size of a marble. That instant is just long enough for a hyperdense piece of plutonium to become hypercritical and experience explosive fission.

It was a brilliant idea, and Oppenheimer made Neddermeyer the head of a new explosives group to thoroughly study the implosion effect. The implosion was simple in concept, but in application it turned out to be extremely complex. Neddermeyer started out with cylindrical shapes, trying to shrink down a rod of metal by putting it in the middle of a cylinder of chemical explosive. The speed with which the shock would develop in the explosive turned out to be very uneven and unpredictable, and his metal rods would end up twisted into odd shapes. After months of unproductive testing, Oppenheimer brought in George Kristiakowsky (1900–82), a Russian-born chemistry professor from Harvard University who was an expert on explosives and chemical kinetics. In mid-June 1944, Oppenheimer read Kristiakowsky's report on the lack of progress in the explosives research, and he made Kristiakowsky head of the group.

Experiments with the U-235 coming in small batches from Oak Ridge indicated that it was better behaved than initially thought, and the length of the Thin Man was reduced to six feet (1.8 m). The uranium bomb using the gun-barrel assembly scheme was renamed "Little Boy." The plutonium-based implosion bomb would be an egg-shaped device, five feet (1.5 m) around and nine feet (2.7 m) long, with fins on the back to make it fly nose down. It was named "Fat Man." By 1945, it looked as though both parallel atomic bomb development paths would result in a practical weapon, and the project raced toward completion.

NUCLEAR WEAPONS RESEARCH IN GERMANY, JAPAN, AND THE SOVIET UNION

The German secrecy structure was as airtight as the Manhattan Project, and there was no word as to what nuclear work was transpiring in the Axis countries. Only after the Allies were able to occupy major territory in Germany in early 1945 were investigators able to evaluate the status of the German atomic bomb effort. A special project, known as the Alsos Mission, was formed by General Groves to capture nuclear personnel, plans, and equipment from the defeated country and discover why no nuclear weapons had been implemented. The project had to work quickly and with maximum priority to beat the Soviets in the last-minute rush to collect German assets.

The Alsos Mission eventually found that the Germans had given up on a full-scale atomic bomb effort in early 1942. The Minister of Armaments and War Production had been killed in a plane crash, and Albert Speer (1905–81), an architect working directly for Adolf Hitler, was named to replace him. Speer immediately took charge of the armaments budget, and in examining the expenditure books he noticed money disappearing into a project labeled "uranium." Curious, Speer arranged a meeting with the principal scientists to find out the nature of this effort. Speer was not impressed. When he asked how long it would take to complete this uranium weapons project, Speer was given the realistic estimate of four or five years. He knew that they did not have four or five years. Germany

After starting off with a firm lead in the race for an atomic bomb, this modest experiment in a cave in Haigerloch, Germany, is all that the Germans had to show for seven years work. Cubes of uranium oxide hanging by wires were lowered into an aluminum pot of heavy water. A self-sustaining fission reaction was never achieved. *(Atomkeller-Museum Haigerloch)*

would run out of fuel to run tanks, airplanes, and trucks in 12 months, and the ability to wage war would come to a stop, so it made no sense to have a war production effort that would take longer than a year to complete. Nuclear research was given appropriately low priority for the remainder of the war.

Imperial Japan seemed less capable of a practical atomic bomb development than Germany, but two such projects were underway when the war ended in 1945. Dr. Yoshio Nishina (1890–1951) established a nuclear research laboratory at the Riken Institute for Physical and Chemical Research in 1931, and by 1937 he had built two cyclotrons, copies of the large machines at Berkeley that were used to transmute uranium-238 into plutonium-239. In 1938, he was able to purchase a new cyclotron from Berkeley. In 1939, the potential for nuclear fission in uranium became clear to physicists all over the world, and in July 1941 Nishina became the director of the Japanese army nuclear program. The mission was to build an atomic bomb for use in conquering territory in the Pacific Ocean.

A small Japanese team managed to build gaseous diffusion apparatus for the critical U-235 separation, but only on a laboratory level and nothing approaching the enormity of the K-25 plant at Oak Ridge, Tennessee. Another limiting problem was the lack of uranium. It was only available through the black market in China and by trade with Germany, but there was no usable transport system between Germany and Japan. Attempts were made to ship uranium to Japan by way of submarine, with no success.

Soviet Russia was ideologically opposed to anything as impractical as nuclear research in the decades leading to World War II, but when nuclear fission was discovered in Germany in 1939 official interest picked up. Most Soviet scientists reasoned that nuclear power production was theoretically possible, but development would take decades. The first work in nuclear research was performed in 1940, confirming that multiple neutrons were released in the debris following a fission of uranium.

By April 1942, it was obvious to the Soviets that the United States had launched a nuclear weapons project because suddenly the American physics journals stopped publishing nuclear research papers. Joseph Stalin (1878–1953), general secretary of the Communist Party, saw this as an ominous development, but the USSR was not in a good position to mount a large-scale scientific industrial project. Stalin chose the next-best option, to thoroughly infiltrate the Manhattan Project with spies.

THE *TRINITY* TEST

The uranium-based atomic bomb design, "Little Boy," was so foolproof there was no question that its use would result in a full nuclear explosion. The only problem was producing the purified U-235, but it looked as though a full bomb load would be ready by August 1945. The configuration was straightforward. A stack of nine rings of U-235, 6.25 inches (15.9 cm) in diameter, were to be shot down a 6.5-inch (16.5-cm) gun barrel using four two-pound (1.8-kg) bags of cordite canon propellant. The four-inch (10-cm) hole in the center of the stack of rings would slide perfectly over a stack of six U-235 disks bolted to the end of the gun and centered in the bore. The stack of rings and the stack of disks would come together quickly as the rings reached the end-of-travel in the gun barrel, slamming into a tungsten-carbide anvil, with the two combined uranium shapes forming the equivalent of four supercritical masses. Instantly, the nose of the bomb would convert into a ball of superheated, radioactive plasma,

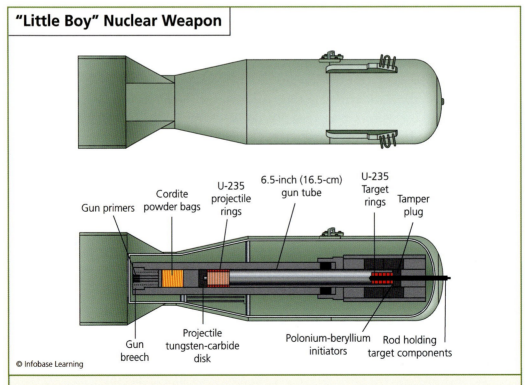

"Little Boy" Nuclear Weapon

Gun primers   Cordite powder bags   U-235 projectile rings   6.5-inch (16.5-cm) gun tube   U-235 Target rings   Tamper plug

Gun breech   Projectile tungsten-carbide disk   Polonium-beryllium initiators   Rod holding target components

© Infobase Learning

"Little Boy" brought two masses of uranium-235 together to form a hypercritical mass, using a gun barrel.

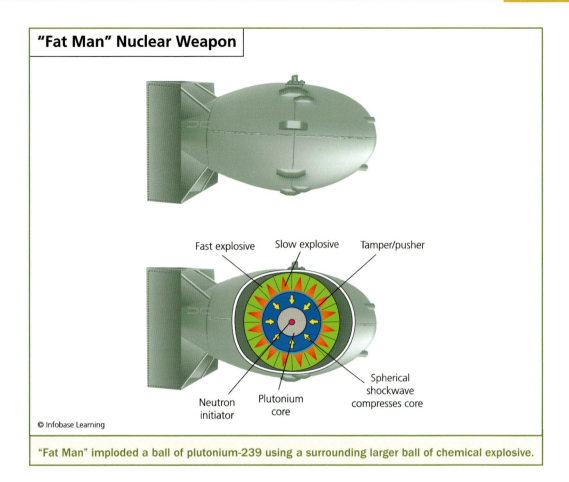

"Fat Man" Nuclear Weapon

Fast explosive   Slow explosive   Tamper/pusher

Spherical shockwave compresses core

Neutron initiator   Plutonium core

© Infobase Learning

"Fat Man" imploded a ball of plutonium-239 using a surrounding larger ball of chemical explosive.

1.25 miles (two km) in radius, as the uranium underwent unimpeded, prompt fission on the high end of the neutron energy spectrum. There was no need to even test this version of the atomic bomb.

The plutonium bomb, Fat Man, to be detonated by the implosion method, was another matter. There were too many unknowns, too many bits and pieces of technology that had never been tried before, and there was too much surplus Pu-239 not to test it. After several configuration changes in early 1945, the final production design, designated Model 1560, was frozen by April 3, 1945. At the center of the 10,265-pound (4,656-kg) bomb was a 14-pound (6.4-kg) ball of plutonium, to be crushed into a hyperdense, hypercritical mass. The ball, or core, 3.62 inches (9.2 cm) in diameter, was surrounded by a shell of U-238, to act as an inertial weight to keep the core together for a few microseconds as it fissioned, surrounded by a thin layer of boron-10 to hold down spontaneous neutrons reflecting back into the core, surrounded by a thick layer of aluminum, to hold it in place. The chemical explosives surrounded the metal core pieces in two layers. The inner layer was 32 close-fitting segments, made of a slow-burning explosive called baratol-70, precision cast like parts of a plastic puzzle. Surrounding the baratol layer were another 32 segments of a fast-burning explosive called Composition B. On the inner surface of each segment of the outer explosive was a depression, backfilled with baratol. These were the explosive "lenses" that would direct the detonation into a shrinking spherical shock wave, imploding the core and starting the fission. Each outer segment of the explosive was equipped with two Model 1773 bridge-wire detonators, all wired to go off at once.

A replica of the bomb to be dropped by airplane over Japan was assembled, having all parts except the outer steel armor plating, and was tested in the desert of New Mexico, early in the morning on July 14, 1945. The test was code-named Trinity, or TR.

The test site was a lonely patch of desert named Alamogordo. On it was erected a 60-foot (18-m) steel tower, bought as surplus from the U.S. Forestry Service, and the bomb was winched into position on a wooden platform at the top. Scientists from the lab were invited to the test, and they could watch from a spot 10 miles (16 km) away from the blast. They were cautioned against looking into the darkness at the point where the explosion was expected to occur, because it was expected to be an unusually bright flash. Most looked in the opposite direction, some just closed their eyes, some looked through welder's goggles, and Richard Feynman (1918–88), an American theorist from MIT and Princeton, decided to sit in a truck and look through the windshield. He reasoned that the ultraviolet rays from the light would be shielded from his eyes by the glass. Oppenheimer, scientific director of the project, watched the 25-minute countdown at the 10-mile point, while General Groves observed from a more discrete 17 miles (27 km). There was a pool of bets on the strength of the blast, ranging from a bet that nothing would happen to a bet that the atmosphere would catch fire and the entire world would be destroyed.

The countdown had to be stopped at 20 minutes, as a rainstorm blew across the test site, with lightning. There was fear that a lightning strike would set off the bomb, but the storm left, and at 5:10 A.M., the countdown resumed. At 5:29:45 A.M., the first atomic bomb exploded. It lit up the darkness like instant noon. Feynman, staring directly at the blast, was temporarily blinded as the brilliant white light overloaded his retinas. The

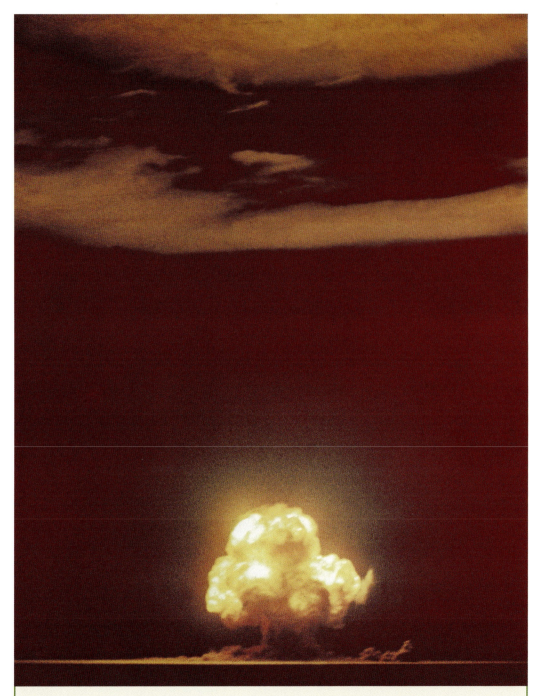

On July 16, 1945, the first nuclear weapon was fired in a test in the desert in New Mexico; it was frighteningly successful. *(Los Alamos National Laboratory Archives)*

light was visible on the horizon 150 miles (241 km) away, and the shock wave rattled windows at a distance of 200 miles (322 km). Performance of the gadget exceeded most expectations, with an energy yield equivalent to 20,000 tons (18,143.7 mt) of TNT, or 84 trillion joules.

JAPAN SURRENDERS

The war was over, but only in theory. The armed forces of Japan were expecting an invasion of the home islands, and they were prepared to repel such intrusion with the lives of every human being who could walk and swing a stick. The Japanese diplomatic corps was counting on a last-minute alliance with the Soviet Union, which would hopefully stop an American ground invasion. On July 28, 1945, the Japanese prime minister Admiral Baron Kantaro Suzuki publicly announced that Japan would ignore the latest peace plan from the Allies, the Potsdam Proclamation, and continue to fight. On August 2, the new president of the United States, Harry S. Truman (1884–1972), weighing the pros and cons of using this new type of weapon, gave the order to drop the bombs on Japan.

The 509th Composite Group of the 313th Bombardment Wing of the U.S. Army Air Force had trained and prepared to drop the atomic bombs. The personnel and aircraft were assigned to the air base on Tinian Island in the Pacific Ocean. Special air-conditioned buildings were erected for assembling and testing the bombs, and loading pits were sunken into the pavement off the runway for gently lifting the heavy devices into waiting aircraft. A new custom-fitted B-29 four-engine bomber, named the *Enola Gay* for its pilot's mother, took off in the early morning of August 6 carrying the uranium bomb, "Little Boy," number L-11.

The target was Hiroshima, the seventh largest city in Japan with a population of about 350,000. It had been spared the bombings of most other industrial cities in Japan, and it was hoped that its complete destruction by a single device would convince the government of Japan of the futility of resistance. Dropping an atomic bomb on Tokyo, the largest city, would have been pointless. All the buildings had long since been burned to the ground or knocked over by a relentless conventional bombing campaign, and an atomic bomb explosion would have made no difference.

The bombing run was perfect and by the textbook, with clear weather, no enemy fighter planes, and no antiaircraft fire. At 9:15:17 A.M. Hiroshima time, "Little Boy" was released from an altitude of 31,000 feet (9,500 m). Exactly 44.4 seconds later, it exploded 1,968 feet (600 m) above the center of the city, and Hiroshima was lost to Japan by a device that had never been tested. Casualties were impossible to count, but are thought to be more than 83,000 people.

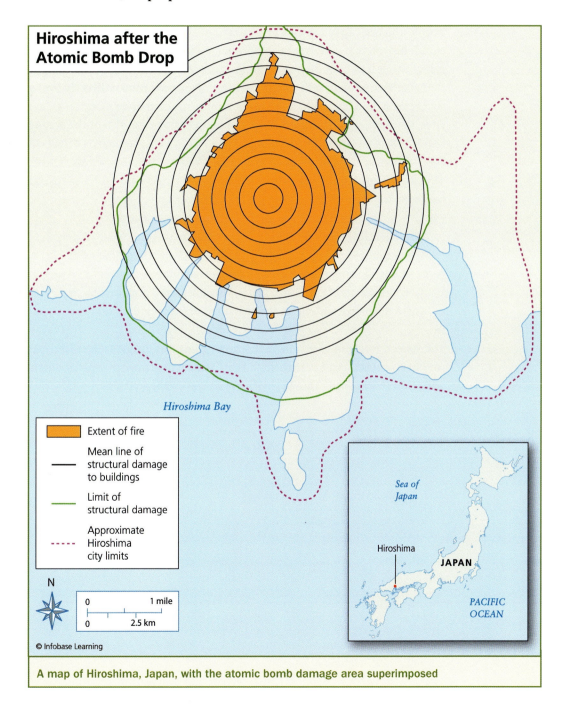

Hiroshima after the Atomic Bomb Drop

Hiroshima Bay

Extent of fire

Mean line of structural damage to buildings

Limit of structural damage

Approximate Hiroshima city limits

N

0   1 mile

0   2.5 km

© Infobase Learning

Sea of Japan

Hiroshima

JAPAN

PACIFIC OCEAN

A map of Hiroshima, Japan, with the atomic bomb damage area superimposed

The government of Japan was oddly silent about this event. It would take a few days for the enormity of it to sink in. Bomber command sent another B-29, "Bock's Car," with a plutonium-fueled implosion bomb on August 9. The target was the undisturbed city of Kokura, home to 110,000 people and the site of a major army arsenal. Fortunately for Kokura, it was clouded over that day, and the bomber moved to the secondary target, Nagasaki. It had 212,000 people and a large Mitsubishi armaments plant.

At 10:58 A.M. Nagasaki time, "Fat Man" dropped, almost directly over a soccer field, and Nagasaki went up in a mushroom-shaped cloud. Six days later, at noon Japan standard time, the people of the Empire of the Rising Sun for the first time in history heard the voice of their emperor, Hirohito (1901–89), over the radio via a phonograph record, made two days before. This unprecedented "Jewel Voice Broadcast" carried a carefully prepared message, beginning as:

> TO THE SUBJECTS OF JAPAN
> After examining Japan's current situation and condition, I have decided to take extraordinary measures. I have ordered our government to inform the governments of the United States, Great Britain, China, and the Soviet Union that Japan will accept the provisions of the joint declaration.

The speech went on to admit that "the war has not progressed entirely as we have wished," and it mentions that "the enemy now possesses a new and terrible weapon." The war with Japan was finally over.

8 Atoms for Peace and Atoms for War

With the end of World War II, the furious push for nuclear development came to a sudden halt. The scientists, engineers, and technicians were tired, and as a group they experienced the depression that can come from the completion of a huge task and the unique dread of having built unusually ferocious weapons. Many workers at Los Alamos, having performed their duty, returned immediately to their lives in academia or industry. Some lingered. Robert Oppenheimer, director of the lab, remained a year and then resigned to return to the Institute for Advanced Studies at Princeton and to become chairman of the general advisory committee for nuclear project funding and laboratory construction. Enrico Fermi, inventor of the nuclear reactor, went back to teach at the University of Chicago.

The Hanford Works in Washington State kept turning out plutonium by the ton, and the gaseous diffusion plant at Oak Ridge, Tennessee, now running at full capacity, started producing bomb-grade U-235, even though there were no plans to build another uranium bomb. The Fat Man design was improved, and a few MK III plutonium implosion devices were assembled, in case of renewed global hostilities. Although building an electrical power plant using the vast wealth of knowledge and experimental data gained during the war seemed a logical idea, it was an idea for the future, and there was no immediate push to civilize the weapons work or to move on to public service applications. In comparison to the effort of building a nuclear-based power plant, a bomb seemed simple. "Little Boy," after all, had exactly one moving part, and it had cost $1.8 billion to build it. A nuclear plant would have turbines, pumps, and valves of all descriptions, electrical controls and monitors, and it would have to be taken apart to refuel it. The machinery would have to be safe enough to run in populated areas without disruptive accidental radiation releases. It was easier to build a nuclear device that would spread radiation than to build a nuclear device that would not spread radiation.

This atmosphere and attitude of not rushing into anything lasted only briefly, and soon there was a push for nuclear power and even a competition among the United States, the Soviet Union, Canada, and Great Britain to tame the wartime technology, and all would eventually claim to have built the first nuclear power plant. The four countries would have different approaches to similar goals, and it is interesting to see the results of the race for atomic power.

THE BUILDING OF THE *NAUTILUS*

Of all the possible applications for nuclear power that tantalized scientists in the early days of development, from nuclear spaceship propulsion to heating an Antarctic station, the most sensible, immediate application was to power a submarine. Submarines in World War II were only marginally capable of submerging for a few hours, operating slowly on batteries or near the surface on diesel engines sucking air through a snorkel sticking above the water. A submarine desperately needed a power source that would require no air and expel no exhaust. With such an engine, a submarine could operate indefinitely under water, and all the military advantages of submersion, stealth, speed, and safety could be used to full advantage. The navy, after an initial recoil from such radically new concepts in 1939, had begun to see the obvious advantages of nuclear power in 1940, but the army's bomb program took supreme priority and command of nuclear matters and confiscated the navy's thermal-column uranium enrichment facility for use at Oak Ridge.

At the end of the war, the U.S. Navy had 1,000 ships sitting idle, quietly rusting away at anchor. The world had just finished a long and costly crisis, and there were other needs to be met. Europe and Japan were in ashes and would need to be rebuilt; Great Britain was starving; and returning soldiers in the United States were jobless. There was also the inconvenient problem of uranium stockpiles. The United States had none to speak of. All the uranium in stock had been bought from Canada and from Belgian companies operating in the Congo, or had been seized in Germany from mines in Czechoslovakia. There was no clear source of uranium for the United States without dealing with touchy international situations. All uranium stocks were frozen, dedicated to future atomic bomb production.

Captain Hyman Rickover (1900–86) of the U.S. Navy was among the first to push for the development of a nuclear-powered submarine. Rickover was ambitious, creative, and tireless. He was also controversial, iron-fisted, and impatient with normal naval channels as a way of doing things. He was a master of the vituperative report, and he drove men and machines to the breaking point. He was famous, in naval circles, as a man who could "get the job done," and he believed that the shortest path was a straight line, even if it cut through several admirals. He had been assigned to a post at Oak Ridge to study nuclear topics, and the concept of a nuclear submarine stuck him as an idea whose time had come. He formed a group of like-minded men at the Oak Ridge Laboratory, "The Naval Group," composed of himself, Lieutenant-Colonel James H. Dunford, Lieutenant-Colonel Miles A. Gilbey, Lieutenant-Colonel Lou Roddis, and Lieutenant Ray Dick. It would require an enormous, seemingly superhuman effort, but this determined group would convince the navy, the Atomic Energy Commission (AEC), Westinghouse, General Electric, the Congress, and the general nuclear physics community that a nuclear submarine should and would be constructed, and it should be a priority project.

The technical problems were small compared to bureaucratic problems, but they were still formidable, and the solution of these problems would affect all future application of nuclear power. At the time, in 1946, a nuclear reactor, or "pile," was assumed to be built using blocks of graphite. Graphite is a fine neutron moderator, having a very small neutron capture probability, and low-grade "natural" uranium can be used as fuel. Graphite is a solid and will neither boil away nor leak. A problem with a traditional graphite pile is that it is huge. The graphite power reactors at the Hanford Works fit in buildings the size of gymnasiums. A submarine, on the other hand, is a slender steel tube, designed to move through water. A graphite pile would simply not fit in even a large submarine, which was only 28 feet (8.5 m) in diameter. The reactor would have to be a completely different design and could not be a modification of a research reactor or a plutonium production facility. An alternate type of moderator material, allowing a small reactor core, would have to be found and proven.

The army, as owner of all the Manhattan Project facilities and every piece of nuclear material and researched knowledge, took a progressive move in April 1946. The Oak Ridge Laboratory was run by the Monsanto Company and the postwar director of the lab was Dr. Farrington Daniels (1889–1972), professor of chemistry at the University of Wisconsin. Daniels proposed that Monsanto, working for the army, build a demonstration industrial pile at Oak Ridge as a prototype civilian nuclear power plant. It became known as the Daniels Pile, and work began immediately, to be completed within 18 months.

Rickover's group watched the Daniels Pile initiative spiral out of control quickly and lose progress. Learning from this observation, the navy team decided to begin anew, designing their naval reactor backward. Instead of starting with the design of the uranium core, they began with the propeller shaft on the submarine. To move a submarine hull underwater faster than a destroyer could run, they needed 10 megawatts of high-temperature steam directed into twin, multistage turbines, turning the two propellers. Another requirement for this reactor was unique: It would be enclosed in a metal tube with a crew of sailors, and they must be able to stay in the tube for an unlimited time without being subjected to harmful radiation. No radioactive substance, such as fuel, *fission products,* radioactive gases, or contaminated coolant could have the slightest leak into the submarine. This was a difficult proposition. Nobody had been killed by the big power reactors at Hanford, so far, but when one was running at full power no one could be near it. Moreover, the reactor had to have an inherently safe character.

The leakage of water from the primary cooling system, such as that caused by the failure of a pump or a valve, would result in a loss of moderator. The dual purpose of the water in the reactor, as both moderator and coolant, meant that losing water was losing moderator. With the moderator gone or reduced in volume, the fission process could not continue. The reactor shuts down and cannot be restarted, and this is considered a safe condition for a nuclear system with broken hardware. This was not necessarily a characteristic of the big graphite reactors. Lose the water coolant through a pipe rupture or boil-off and the reactor would go supercritical because the graphite alone was a better neutron moderator than graphite with water running through it.

As an additional requirement, the reactor had to be able to run for years at full speed without refueling, unlike the graphite piles. Using natural uranium, the fuel in a graphite reactor had to be almost constantly replenished, pushing new fuel in one side of the pile with spent fuel falling out the other side of the pile, on a weekly basis.

Working in reverse, the submarine design continued from the turbines to the steam generator, with sizes and weights determined by the volume and pressure of steam needed to provide the specified power. The steam generator, or boiler, needed a source of heat. The source of heat was the combination coolant and moderator for the reactor core.

Instead of assuming that there would be no uranium fuel, Rickover assumed that there would be plenty of fuel. The United States is a big, wide country, and there had yet to be a comprehensive survey of available uranium deposits. Given unlimited uranium reserves, a submarine could run on enriched fuel, made 50 percent U-235 in the diffusion plant at Oak Ridge. Given the high concentration of fissile U-235, graphite or any other high-efficiency moderator was not necessary. Ordinary or "light" water would be sufficient. The neutron slowing-down distance for the hydrogen in ordinary water is unusually short, so the core could be very small, the size of a garbage can. Water was well understood, easily pumped and managed, transparent, and liquid under pressure, regardless of the temperature. Liquid water under pressure, pumped in a tight loop through a reactor core, is both the coolant and the moderator. Lose coolant through mishap, and the reactor shuts down, because moderator is lost. Any fission-products or broken fuel are confined to this tight, inner loop of water, so everything on the other end of the power plant, from the steam generator to the turbines and the condenser, is isolated and free of potential radiation contamination.

One thing that Rickover had learned from his service on creaky, prewar undersea vessels was how not to design a submarine, and his plan, now named *Nautilus* after the craft in Jules Verne's *Twenty Thousand Leagues Under the Sea,* would be safe and solid. The reactor plant would be inherently foolproof, free from possible criticality excursions, with far more strength than was necessary in all important components. Every piece of technical equipment on Rickover's submarine would be built to shock and vibration specifications, with verification through physical testing.

Official approval for his submarine project in the navy would require the highest official approval, and Rickover decided to shorten the tedious chain of command, writing a letter directly to Fleet Admiral Chester Nimitz (1885–1966), decorated hero of World War II, former commander in chief of the Pacific Ocean areas, now chief of naval operations, and the navy's principal expert on submarines. Writing a letter to Nimitz was hardly a trivial matter, and even though directly addressed it still required approval and rewriting all up the line, from Rickover's captain level up to fleet admiral level. It took Rickover two months to build the letter and send it through official channels. Finally, on December 5, 1947, the letter reached Admiral Nimitz.

Nimitz was fascinated by the details of the proposed project. He signed the letter immediately, approving the program to build a nuclear-powered submarine. Still, much work was needed to persuade both the navy and the AEC that *Nautilus* should be built, but on May 1, 1948, the concept had full approval. On August 2, Rickover formed the nuclear power division of the Bureau of Ships and re-collected his naval group from Oak Ridge. The Westinghouse Corporation was chosen to design a most critical part of the system, the steam generator, under an $830,000 contract.

Technical problems in *Nautilus* were interesting and numerous, and their solutions would forever guide the nuclear power industry. An example was the production of zirconium. It was found early on that zirconium was ideal for internal structures of a submarine reactor. It could withstand high temperatures and did not absorb neutrons. Unfortunately, it was a rare material, more precious than platinum. They would need a lot of it for one reactor. A metallurgist broke the news of its cost to Rickover. It would be more than $1,000 per gram.

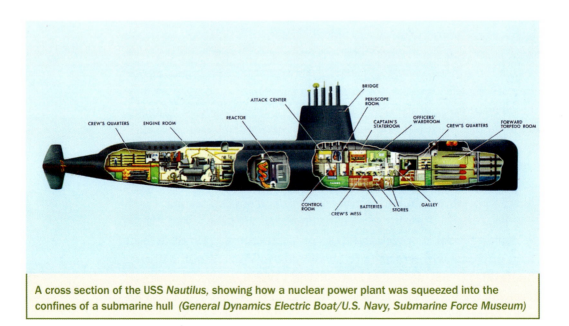

A cross section of the USS *Nautilus,* showing how a nuclear power plant was squeezed into the confines of a submarine hull *(General Dynamics Electric Boat/U.S. Navy, Submarine Force Museum)*

"My God," Rickover said, "$1,000 a gram is $450,000 a pound. A half a million dollars a pound! . . . What's the problem?"

The problem, explained the metallurgist, was that all the zirconium in the United States could be put in a shoe box.

"Well, we've got to step this thing up," replied Rickover. "From now on you call me Mr. Zirconium, because I am going to get this stuff produced by the ton."

And so he did. By 1952, zirconium was being mined, milled, and produced in quantity and at low price, by Westinghouse. Asked by a congressional committee how in the world they managed to get the machinery and the expertise to make zirconium so quickly, the Westinghouse representative replied, "Rickover made us get it." The zirconium exercise almost made the procurement of hafnium, a rare metal resistant to high temperature and perfect for the *Nautilus* control rods, seem simple.

With technical issues resolved, the first submarine reactor was built in 1952 by Westinghouse and the Electric Boat Company of Groton, Connecticut, but it was nowhere near the ocean. It was built in a simulated ocean, a spherical building 18 stories tall, 225 feet (69 m) in diameter, filled with seawater and named the "Hortonsphere." Located in the wilds of Idaho, near the town of Arco at the AEC's Desert Test Station, the construction was a tightly held military secret. Those of little faith believed that an experimental reactor should be built where it could explode harmlessly, and Rickover believed that it should be proven in a simulated submarine hull, completely under water.

The development of Rickover's submarine reactor would have a profound effect on future nuclear power plant design in the United States. As the state of nuclear power production stands today, most of the reactors in the United States and in the world are based on the equipment in Hyman Rickover's submarine, the *Nautilus*. The choice of moderator and coolant, reactor tank configuration, piping, safety systems, controls and monitors, exotic materials used, and even operator training are all directly connected to the navy's first nuclear submarine program. The result is compact machinery, overbuilt for stamina, expensive, highly reliable, and safe to operate. The nuclear power program for the United States and most of the world was thus designed in reverse, starting with a highly specialized system, built to fit in a small space, with price being no object and using enriched uranium fuel, instead of first building a plant spread out over many acres, using readily available machinery and materials and the most inexpensive form of fuel. There would be other designs, but for better or

The world's first nuclear-powered submarine, the USS *Nautilus*, making a high-speed turn on the surface, where it spent very little time. *(U.S. Navy, Submarine Force Museum)*

worse, this is the system that Rickover gave the nuclear industry. The navy reactor is known as the pressurized water reactor, or the *PWR*. It is the most licensed, copied, and stolen reactor plant design in the world.

The keel of the *Nautilus,* hull number SSN-571, was laid on June 14, 1952, at the shipyard in Groton, Connecticut. The president of the United States, Harry S. Truman, was present, as were the secretaries and chiefs of the armed forces, the governor of Connecticut, the chairman of the AEC, lesser officials of all types, and Hyman G. Rickover. The radical new vessel was launched into the Thames River on January 21, 1954. It took 10 months to install the nuclear power equipment, and at 11:00 A.M. on January 17, 1955, she put to sea under Captain Eugene P. Wilkinson (1918–   ).

The age when uranium fission would be used for something other than explosives had begun.

THE ATOMIC ENERGY ACT AND ATOMS FOR PEACE

The United States was not the only country with nuclear power ambitions. The Union of Soviet Socialist Republics, under the leadership of Stalin, built a new science city, Obninsk, 68 miles (110 km) southwest of the capital, Moscow. On January 1, 1951, construction began in Obninsk on the AM-1 nuclear power station. It was not a highly innovative design, as it used the reactor configuration and material choices from the wartime plutonium reactors built in Hanford, Washington, from information gained by expertly executed espionage. First startup was on June 1, 1954, coming up to full power and connecting to the electrical power grid 25 days later. It produced only five megawatts of electricity, hardly enough for half the town of Obninsk, but it was still the world's first civilian power station.

With its graphite moderator combined with water cooling and its use of inexpensive, natural uranium fuel, the reactor was not designed for optimum safety, but it would be a prototype for more ambitious Soviet plans. Its design would be scaled up into large, practical power-production reactors, each generating 1,500 megawatts of electricity, named the RBMK, reaktor bolshoy moshchnosti kanalniy, or the channel-type high-power reactor. The explosion and *meltdown* of an *RBMK* near the town of *Chernobyl* on April 26, 1986, would change the course of nuclear power development worldwide, as the severe radiation release led to the permanent evacuation of the town and its surrounding area. The Chernobyl incident involving RBMK-4 would be the worst nuclear accident in history.

Great Britain after World War II had the advantage of having participated with the United States in the intense wartime nuclear development, and British scientists returned home with ideas and plans for making their country part of the new atomic age. Although in 1946 the United States had closed its nuclear programs to all other countries, the British government started an independent weapons and power program at the *Windscale* facility, near the village of Seascale, on the Cumbrian coast. Their first nuclear reactors were two plutonium-production units, Windscale Piles 1 and 2. In retrospect, these were probably the least safe reactors ever built, with less safety margin than the Soviet AM-1. Unlike Rickover's

ADMIRAL HYMAN RICKOVER: FATHER OF THE NUCLEAR NAVY

Hyman George Rickover was born in 1900 in the all-Jewish village of Makover in Russian-occupied Poland. His father, Abraham, was a tailor, and the Rickovers were the poorest of the poor. The future did not look bright. The expressed objective of the Russian government was to convert one-third of the Polish Jews to the Russian Orthodox religion, force one-third to leave, and kill the rest. The Rickovers chose the second option, working west across Germany in 1905 and finding passage to New York City at Antwerp, Belgium. Young Rickover, the future admiral, reportedly burst into tears at the sight of an oceangoing ship. His older sister, Fanny, reports that, "The boats were so big, they frightened him."

The family landed in New York and moved west to Chicago. Rickover added income to his father's tailoring work beginning at the age of 9. He would later sum up his childhood as a time of "hard work, discipline, and a decided lack of good times." While attending John Marshall High School in Chicago, Rickover had a full-time job delivering Western Union telegrams by bicycle. By fate, he was appointed a Western Union messenger at the Republican National Convention in Chicago in 1916, where he met Congressman Adolph Sabath, a fellow Jewish immigrant from Czechoslovakia.

naval reactor design, the Windscale units had to use natural uranium in metallic form. No other country had the luxury of large industrial-scale uranium enrichment, and early reactor designs in other countries seemed to all rely on natural fuel. Rickover's naval reactor design also used uranium fuel in oxidized form, considered advantageous because of its high melting point. The Windscale had to use pure uranium metal, to maximize the poor reactivity of its unenriched uranium. Unfortunately, uranium metal has a low melting point, and it will catch fire in air and burn like gasoline.

A Windscale reactor simply pulled air in one side of the building with a fan, blew it through holes in the pile of graphite and fuel, and sent it up a smokestack, to cool the reactor as it converted U-238 into Pu-239. A large fiberglass filter at the top of the stack was supposed to trap fission products that could escape the reactor core. As might be predicted, on

In 1918, Sabath nominated Rickover for appointment to the U.S. Naval Academy, and he won a coveted place in the class of 1922.

Never known as one to enjoy a party, Rickover quickly gained a reputation as a silent scholar, learning to stay up long after lights-out for studying. Promptly after reporting to the academy, Rickover came down with diphtheria and missed several weeks of introductory instruction. He struggled to overcome his late start by avoiding weekends and sociable get-togethers as he withdrew to the privacy of his room. He was academically above the mean, but his abysmally poor military drilling skills pulled down the class average.

After additional education in submarine school, Rickover was assigned service in the submersible boat S-48. It was a cramped, damp working environment, and he and the crew seemed in constant danger from the failure of unreliable equipment. A sister submarine, the S-49, had recently suffered an explosion of the storage battery, and the S-5 and S-51 boats had sunk with all hands for reasons unknown. U.S. Navy submarines were risky transportation at the time, and within 40 miles of leaving port, the battery in Rickover's boat, the S-48, caught fire while running on the surface.

Rickover was able to evacuate the crew to the narrow, wooden deck. He went back down into the smoke-filled boat to the battery hold, found the fire, and extinguished it. From that moment on, he decided to make submarines the safest and most prestigious service in the armed forces. In time, he would succeed in glorious fashion.

October 1, 1957, the Windscale Pile-1 was running hot and the fuel caught fire. Released into the surrounding countryside were 20,000 *curies* (750 tbq) of radiation up the smokestack, making it the second worst reactor accident in history. It took a while, but eventually the graphite moderator also caught fire as the fuel glowed bright red. Inadequate instrumentation and cost-saving construction measures were blamed. The British nuclear establishment, shaken by the Windscale experience, improved its engineering and built a carbon dioxide–cooled graphite reactor named Calder Hall. This reactor produced 50 megawatts of electricity, as well as plutonium for bombs, and it went online on August 27, 1956. It was the world's first commercial power reactor.

Canada started its own nuclear program immediately after World War II, using knowledge gained by Canadian scientists on temporary loan to the Manhattan Project. An impressive research complex was built at

Chalk River, Ontario, and a 10-megawatt research reactor came online in 1947. It was named the *NRX,* or the nuclear reactor experiment, and it would have the dubious distinction of experiencing the world's first core meltdown on December 12, 1952. The Canadian reactor at least did not use graphite as the moderator, but it did use natural uranium fuel, and to achieve criticality with the low U-235 content a highly efficient moderator was needed. Heavy water, produced in Canada, was chosen, with ordinary or *light water* run through tubes in the reactor core to act as coolant.

The combination of heavy water moderation and light water cooling was a weakness in the design, as a loss of light water by boiling or leakage would actually improve the use of neutrons, causing the power level to increase out of control. Through a series of unfortunate errors, the power level ran away, melting a portion of the reactor's metallic uranium fuel. Several months of cleanup were required to remove radioactive contamination from the failed fuel, and the U.S. Navy sent 150 nuclear-trained personnel to help. Among the workers sent was Ensign James Earl Carter (1924–   ), future president of the United States during the most critical period of nuclear power development. Ensign Carter developed a cautious attitude toward nuclear power during his duty in Canada.

The hidden advantages of Admiral Rickover's unrealistically expensive excursion into nuclear power production, using exotic materials, relentless testing, and a cost-is-no-object attitude began to make grudging sense in the nuclear world. To approach a nuclear power economy from a sensible, businesslike, cost-saving direction was possibly not the best strategy.

Rickover's submarine was a nuclear-powered weapons platform, armed with the latest undersea torpedoes, but this warship was not built by the armed forces. Technically, it was built and owned by a civilian agency, the AEC, and to administer the project Rickover had to be a member of this new organization. The Congress, after months of intense debate among politicians, military planners, and nuclear scientists, presented the McMahon Act to President Truman. He signed it into law on August 1, 1946. This law transferred ownership of all nuclear weapons, uranium stockpiles, and research facilities from the army to the AEC. David Lilienthal (1899–1981), former head of the Tennessee Valley Authority and a capable public servant, was named chairman.

Being head of the AEC was rough duty in the early days, and chairmen changed every few years. Bare-knuckle politics, communist infiltration, Soviet bomb-making, and controversial regulatory stances took their toll. The existing nuclear labs, hastily built during the war, were reorganized into a network of national laboratories. The first formed was Argonne National Laboratory, created near Chicago by Enrico Fermi's CP-1 project. The Oak Ridge Laboratory, Los Alamos, and Hanford soon followed, and additional laboratories would be built all over the country.

The AEC made itself the only legal buyer of uranium, and by artificially setting the price high hoped to increase the incentive for prospecting, particularly in the old vanadium mining territory in the Southwest where uranium could be extracted as a by-product. By December 1949, there was a uranium rush, begun by an article in the *Engineering and Mining Journal* detailing new AEC bonuses. Miners, geologists, and prospectors bought radiation counters and began sweeping the Four Corners area of the Colorado Plateau. One particularly impoverished prospector, Charles A. Steen (1919–2006), unable to afford a Geiger counter, made a massive find using unconventional geological theories. He made claim to a huge, concentrated uranium deposit in the Big Indian Wash of Lisbon Valley, southwest of Moab, Utah. His single "Mi Vida" mine would produce all the uranium the AEC could use for the next decade. To replace his tar-paper shack, Steen built a $250,000 mansion overlooking the works.

On December 8, 1953, President Dwight D. Eisenhower (1890–1969) delivered a speech at the United Nations General Assembly in New York City. It is known as the "Atoms for Peace" speech, and it was the tipping point for international focus on the potential uses for nuclear power in a world without war. It was a profound and noble speech, and a quotation from it follows:

> To the making of these fateful decisions, the United States pledges before you—and therefore before the world—its determination to help solve the fearful atomic dilemma—to devote its entire heart and mind to find the way by which the miraculous inventiveness of man shall not be dedicated to his death, but consecrated to his life.

Hoping to avoid a protracted nuclear weapons race with the Soviet Union, Eisenhower tried to emphasize the possible beneficial uses of nuclear energy, with particular interest in electrical power. To back up his words, Eisenhower then launched a declassification effort, taking any documents sealed during the Manhattan Project that were not directly related to bomb production out of secret status and making the information available worldwide.

THE BORAX REACTORS IN IDAHO

In 1949, the lack of uranium fuel resources in the United States was considered serious, so the earliest government-funded nuclear power project was to build a plutonium-fueled *breeder reactor*. Walter Zinn (1906–2000), the Canadian-born head of the Argonne National Laboratory, led the design of the first power reactor, designated Experimental Breeder Reactor One, or *EBR*-1. Dr. Zinn had been in nuclear engineering from the very beginning, having worked with Leó Szilárd in 1939. He had manned a control rod at Fermi's CP-1 experiment in Chicago in 1942.

EBR-1 was built in the barren desert of Idaho, 18 miles (29 km) southeast of Arco, and it first achieved nuclear criticality on August 24, 1951. On December 20, the secondary steam-coolant loop was switched into a turbogenerator, and electrical power was generated for the first time using nuclear power. Enough electricity was generated to light exactly four bulbs. This modest triumph was added to in 1953, when analysis of the fuel proved that it was making more plutonium than it burned, thus proving the breeder hypothesis. On November 29, 1955, a slight operator error led to a partial core meltdown, which seemed the fate of a few early reactors. EBR-1 seemed unstable at high coolant flow rates, as the rushing liquid metal coolant would buckle the fuel rods and cause the power to rise. It was not the behavior desired in a power source.

In 1952, Samuel Untermyer II (1912–2001), a mechanical engineer from MIT who had worked with Walter Zinn at Argonne, theorized a bold simplification of Rickover's compact pressurized water reactor. The PWR required two coolant loops. An inner loop ran water through the reactor core, transferring out the power and keeping the core at well below melting temperature. The water was under pressure, so that it would not boil. The secondary loop of water ran through a heat exchanger, or steam generator, transferring power out of the inner loop and into turbogenerators. Untermyer suggested eliminating the inner loop, boiling the cooling water to steam in the reactor core, instead of in an external steam generator. This would eliminate much plumbing and pumping, and it appealed very much to the mechanical engineers.

Nuclear physicists were skeptical, arguing that a chaotic boiling of water in the core would result in unpredictable controls, spoiling one of the great advantages to light water moderation and cooling. Untermyer promised that just the opposite was true. The fact that the moderator was allowed to boil in the reactor core would lead to greater stability, and not less. If a steam bubble develops in a reactor core, then moderator is displaced out of the core. With less moderation, the average neutron energy goes up, and the fission rate goes down. Power drops. If the water becomes too cool, then bubbles stop forming. The density of the water increases, and the moderating quality of the water improves. The fission rate climbs. Power goes up. Not only would a boiling water reactor, or *BWR*, be stable, it would be self-controlling, with no need to constantly adjust neutron-absorbing control rods to keep the power level steady. Lose coolant catastrophically in an accident, such as a pipe rupture, and the reactor shuts down instantly, just as in a PWR being designed for the navy.

The *BORAX*-1 experimental boiling water reactor was built in 1953, and the experiments had to be performed in the summer, because there was no building housing the reactor and snow would cover it in the winter. The core tank was semi-buried, four feet (1.2 m) in diameter and 13 feet (4 m) high. The small reactor performed perfectly, proving Untermyer's hypothesis. In 70 experiments, the core was put through severe runaway situations that would have surely melted a lesser design and proved that the BWR provided inherent protection from water-loss hazards. As a final ultimate test, the scientists decided to subject the reactor to the worst possible accident situation before dismantling it. They rigged the control rods to blow out the top of the core, subjecting

A cutaway diagram of the BORAX-V boiling water reactor test facility. The five BORAX reactor setups in Idaho proved the utility and safety of the boiling water concept. *(Argonne National Laboratory)*

the reactor to prompt supercriticality. There was nothing worse that could happen to a nuclear device.

The results of this experiment were somewhat larger than predicted. The instant power increase and the resulting steam explosion carried away the entire control rod mechanism, weighing 2,200 pounds (998 kg), and released 135 megajoules of energy, the equivalent of 70 pounds (32 kg) of high explosive. Pieces of the uranium fuel were found up to 200 feet (61 m) away. Still, the reactor was no longer generating power immediately after it had exploded. This was the worst that could happen, and the BWR concept had proven to be a highly advantageous reactor design.

In 1954, BORAX-II was built, this time having a tin building over it and producing 6.4 megawatts of steam. It proved an important point, that fuel contamination in the single coolant loop would not be a problem, as had been suggested, and in 1955 a turbogenerator was added. For an hour on July 17, 1955, it was connected to the local power grid, and it provided electricity for the city of Arco, the entire BORAX test facility, and part of the National Reactor Test Station. Arco, Idaho, thus became the first city in the world to be powered solely by nuclear energy, and the experiment became BORAX-III.

Samuel Untermyer was the sole inventor of the boiling water reactor, and he was granted U.S. patent number 2,936,273 in 1960. Today there are 90 boiling water reactor plants operating in 10 different countries. It is possibly the safest power reactor design in the world.

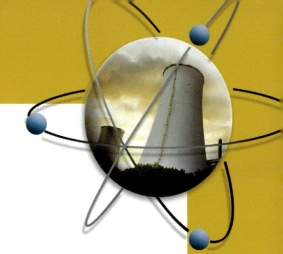

9 America Goes Nuclear

In 1953, Lewis L. Strauss (1896–1974), a retired rear admiral in the U.S. Navy, was named head of the Atomic Energy Commission (AEC). It was an optimistic time, with the world experiencing peace, stability, and rising prosperity, and there was hope and expectation that the secretive technology that had been developed during the atomic bomb project would be put to good and practical use. In 1954, the U.S. Congress passed amendments to the Atomic Energy Act of 1946, freeing information and technology held by the military and making it possible to develop commercial nuclear power operations. By this amendment, the AEC was assigned the dual role of encouraging the use of nuclear power in the civilian sector and monitoring and regulating its use to ensure public safety.

The nuclear power industry did not exist in 1954, and the results of the past 15 years of intense research were mostly locked under military secrecy. Safety regulations and measures were made up as the situation demanded. Two men had already been killed in criticality accidents at the Los Alamos Laboratory, in which masses of fissile plutonium had been carelessly assembled by hand, resulting in lethal flashes of extreme radiation. A reactor core had melted at Chalk River, Canada, because of poor controls and procedures, and then the Experimental Breeder Reactor experienced a meltdown in Idaho. The double assignment of the AEC of pushing forward a new, unknown technology as quickly as possible while imposing strict but nonexistent safety standards would pose an interesting set of problems. The technology to be developed was both ultimately powerful and inherently dangerous. Simply stated, operators of industrial equipment will make errors. If nuclear power was to be part of the commercial power industry, then an elementary operator error could lead to equipment damage and injuries, but it could not lead to a melting of the capital equipment and the evacuation of an entire city. The next 25 years would involve much learning.

Strauss stepped vigorously into the role of AEC chairman, and on September 16, 1954, he gave an important speech at a meeting of the National Association of Science Writers. Speaking of the coming age of nuclear power, he said the following:

> It is not too much to expect that our children will enjoy in their homes electrical power too cheap to meter; will know of great periodic regional famines in the world only as matters of history; will travel effortlessly over the seas and under them and through the air with a minimum of danger and at great speeds, and will experience a lifespan far longer than ours, as disease yields and man comes to understand what causes him to age. This is the forecast of an age of peace.

His prediction of "electrical power too cheap to meter" would haunt the nuclear power industry for decades to come. It was true that using uranium fission electrical power could be made at a rate so inexpensive that a power meter on each house would be a superfluous waste of money. The volume of uranium that would be fissioned for every person in the United States per year for electrical power needs was miniscule, compared with the stockpiles amassed and the potential uranium ore in the ground. However, this prediction assumed that fuel would be used in its cheapest form, and that power plants would be built in the least expensive way possible, just as power plants had been built since the beginning of electrical power usage. The United States and the world would come to learn that nuclear fission was a new way to generate power in more ways than one. The least expensive option would apply no longer.

In this chapter, the maturing process in the nuclear industry and in the world's perception of nuclear power are examined in chronological order, as new power plants were built and tested on the ascending node, and as major accidents occurred on the descending node. Commercial nuclear power is shown making its debut and its rise in industrial popularity and then its decline as the bottom line of profitability becomes evident. In the final topic of this chapter, the needs of a maturing world economy with respect to nuclear power are examined, as atmospheric chemistry and limited burnable resources become important considerations for power generation. A sidebar details an important federal requirement for nuclear power plant construction, the Final Safety Analysis Report.

THE FIRST CIVILIAN POWER REACTORS

At the end of World War II, the U.S. government upgraded the existing, hastily constructed nuclear laboratories at Los Alamos, New Mexico, and Oak Ridge, Tennessee, and built several more across the country. Major labs were built in Brookhaven, New York, Livermore, California, Idaho Falls, Idaho, Aiken, South Carolina, and at Simi Hills, California.

At Simi Hills, overlooking the Simi Valley 30 miles (48 km) north of Los Angeles, the *Santa Susana Field Laboratory* (SSFL) was constructed on 2,668 acres (10.8 km²) of land. Its purpose was to test rocket engines, guided missiles, munitions, and nuclear reactors in what was then a sparsely populated area. As seemed usually the case in the 1950s, nuclear engineers were interested in exotic reactor designs using high-performance coolants, such as liquid sodium. A commercial nuclear power plant was developed at the SSFL. It was named the Sodium Reactor Experiment.

Sodium has some good properties as a reactor coolant, but it also has a few disadvantages. It does not absorb neutrons parasitically, nor does it boil away at anything but the highest temperatures, and it has excellent heat conduction. However, at low temperature it is a solid, locking everything immersed in it in a metallic block. It is opaque. You cannot simply look and see what is going on in a sodium-cooled reactor core. It also reacts vigorously with air or water vapor, and this means that it cannot be allowed to leak out into a room. Its compound with water is extremely corrosive and will quickly dissolve aluminum.

The sodium reactor experiment was brought to power operation in April 1957, and on July 12, 1957, its electrical output was switched into the California power grid, making it the first commercial nuclear power production reactor in the United States. For a short period of time, just long enough to prove the point, it supplied power to 1,100 homes in the Moorpark area of California. On July 13, 1959, the Sodium Reactor Experiment made another first. It became the first power-producing nuclear reactor in the United States to experience a core overheating.

The reactor was operating normally when it experienced a sudden power excursion, with the power level and temperature rising rapidly. With considerable effort, the reactor was brought under control and shut down. The cause of the excursion was baffling and was not determined, but the decision was made to ignore the problem as an unexplained anomaly and continue operating as if nothing had happened, so a few hours later the reactor was restarted and taken to operating power.

Subsequent reactor behavior seemed strange, and radiation alarms kept going off, so after 13 more days of wrestling with the controls the reactor was shut down for analysis. The operating crew discovered that almost one-third of the reactor core had melted, releasing radioactive fission products into the liquid metal coolant. Radioactive gases from the wrecked core were collected in holding tanks and then bled into the atmosphere over a period of several weeks. The problem had been caused by leaking seals in coolant pumps. When the seals failed, the coolant for the pump bearings leaked into the sodium coolant. The coolant was an exotic organic fluid, Tetralin, and it carbonized when it hit the sodium, blocking coolant passages. The blockage kept coolant from the fuel, and the cladding melted in the increased temperature. The reactivity of the graphite-moderated core improved without coolant, causing the power level to rise out of normal control.

There was a weakness in the reactor design, and a simple, predictable problem led to a major breakdown. Pump-seal coolant will eventually leak, as the moving parts experience wear. Any flow disruption of the molten sodium in the core would lead to a runaway, instead of an automatic, shutdown. A stronger design would allow anything that is capable of failure to fail without causing a larger problem, and a disturbance of the coolant should cause the reactor to revert to a safer condition and not a less safe condition. In retrospect, the operating procedures for the reactor were fundamentally wrong. When a problem with unknown characteristics arises, the reactor should not be restarted until the cause is known. These were the lessons learned from the Sodium Reactor Experiment meltdown. These lessons would require further reinforcement, but it was a beginning of the nuclear power learning process.

In parallel to the sodium reactor development at the SSFL, Admiral Hyman Rickover oversaw the naval reactors program for the AEC. His development of the power plant for the nuclear submarine *Nautilus* proved remarkably successful. Based on his pressurized water design,

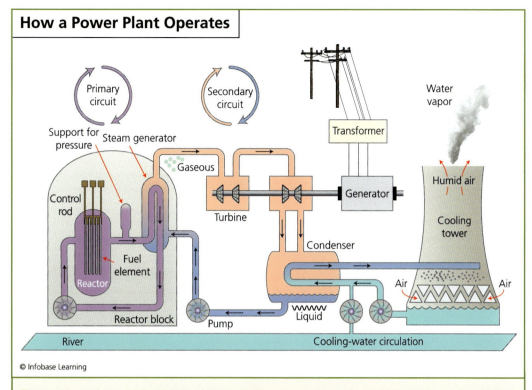

How a Power Plant Operates

The workings of a typical pressurized water nuclear power plant, showing the reactor and its closed coolant loop inside the containment structure, the outside steam loop that turns the generator, and the third water loop that dumps excess heat into the cooling tower

using ordinary water for both the reactor coolant and the neutron moderator, the naval reactors amassed a perfect safety record. No reactor in the U.S. Navy has ever experienced an accidental release of radioactive material. In Rickover's nuclear navy there has never been a reactor meltdown. Never has a sailor or the environment surrounding a nuclear-powered navy vessel been subjected to abnormal radiation due to a malfunction.

The navy, greatly pleased with the nuclear submarine program, ordered more submarines and nuclear-powered surface ships. The 10-megawatt Westinghouse reactor used in the *Nautilus* was upgraded into a 60-megawatt design for use in an ambitious Navy plan to build a nuclear-powered aircraft carrier. Under the auspices of Hyman Rickover in his role inside the AEC, the aircraft carrier engine design was modified for use in a stationary, full-scale electric power plant, in a project proposed by the Duquesne Light Company.

On the Ohio River in Beaver County, Pennsylvania, about 25 miles (40 km) from Pittsburgh, the Shippingport Atomic Power Station was built, beginning on September 6, 1954. It was the cornerstone of President

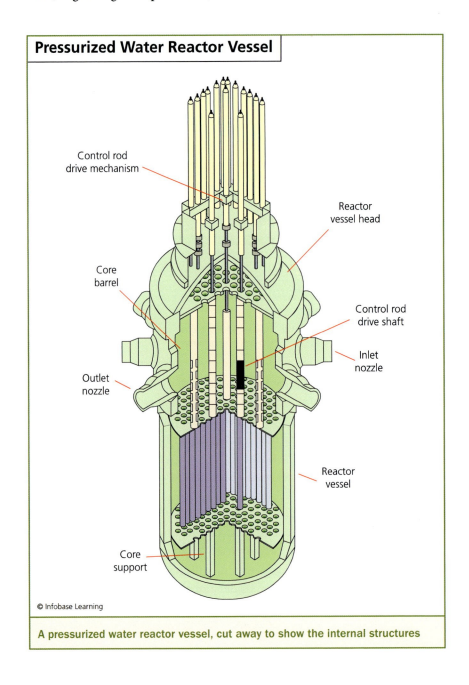

Pressurized Water Reactor Vessel

Control rod drive mechanism

Reactor vessel head

Core barrel

Control rod drive shaft

Outlet nozzle

Inlet nozzle

Reactor vessel

Core support

© Infobase Learning

A pressurized water reactor vessel, cut away to show the internal structures

The Shippingport Atomic Power Station in eastern Pennsylvania. This first civilian nuclear power plant was built around a navy aircraft carrier reactor. *(U.S. Department of Energy)*

Eisenhower's Atoms for Peace concept, and he turned the first shovelful of dirt at the groundbreaking ceremony. It cost only $72.5 million to build, because all the expensive up-front engineering had been paid for by the U.S. Navy.

It took 32 months to build the plant, and the reactor first started up at 4:30 A.M. on December 12, 1957. The plant was brought to full power 21 days later, after the correct operation of all systems had been checked and confirmed. After May 26, 1958, Shippingport was online and officially generating power. It was the world's first full-scale atomic power plant devoted exclusively to peacetime uses. It generated electricity without a problem for 25 years, and it seemed to prove that nuclear power could be used safely and that it was more economical than a conventional plant. There was no need to constantly move train-cars of coal on and off site, and there was no smokestack pouring soot and carbon

SAFETY ANALYSIS

The construction, ownership, and operation of civilian nuclear power plants are conducted under permanent rules and regulations set out in the Code of Federal Regulations and published in the *Federal Register.*

Nuclear power is regulated under Title 10 of the Code of Federal Regulations, which covers all issues of energy. Part 50 of Title 10 covers "Domestic Licensing of Production and Utilization Facilities," and Section 34 of Part 50 is concerned with "Contents of Applications; Technical Information." Paragraph (a) of Section 34, or 10 CFR 50.34(a), states that an application for a nuclear power plant construction permit must include a preliminary safety analysis report (PSAR), and 10 CFR 50.34(b) further specifies that an application for an operator's license for a nuclear plant must include a final safety analysis report (FSAR).

The PSAR is an accounting of the engineering and operating procedures of a nuclear power plant pertaining to safety features or the handling of emergencies. This specification covers 12 points of nuclear power plant safety:

☢ A description and safety assessment of the site on which the plant is to be located, the intended power level and an inventory of on-site radioactive materials, a description of unique or unusual safety features of the facility, a description of radiation barriers that will be in place, and an assurance that an individual standing at the outer fence will not receive a hazardous dose of radiation in the event of the worst possible accident.

dioxide into the air. It had a cooling tower, just as any steam-operated plant would have, whether the heat was generated by burning coal or nuclear fission.

NUCLEAR POWER BECOMES COMMERCIAL

Construction was begun in 1953 on the Calder Hall nuclear plant at *Sellafield* in Great Britain, and it proved to be a highly reliable power source. It was first connected to the power grid on August 27, 1956, and the plant was formally opened by Queen Elizabeth II on October 17, 1956. When it finally closed down on March 31, 2003, the first nuclear plant to

❖ A summary description of the plant, stressing unusual features and safety considerations.

❖ The preliminary design of the plant.

❖ An analysis of the design and performance of the structures, systems, and components of the plant, so that the risk to public health can be assessed. Emphasis is given to the emergency core cooling system, or ECCS.

❖ An identification of items that are of particular interest for the evaluation of the safety of the plant.

❖ A plan for the organization of the plant, the training of personnel, and rules for the conduct of operations.

❖ A description of the quality assurance program.

❖ An identification of any features of the plant that may require research and development.

❖ The technical qualifications of the applicant, or the organization applying to build a nuclear plant.

❖ A discussion of plans for dealing with emergencies.

❖ A discussion of possible hazards to the structures or components of the plant due to the construction features, and administrative controls that will be in effect during construction.

❖ Assurance that the plant will be built to withstand an earthquake.

Filing a PSAR is the first step in the paperwork necessary to build a nuclear power plant. From there, the procedure gets complicated.

deliver commercial power had been in constant use for 47 years without incident. Although it generated power, the Calder Hall reactor's first intention was to produce plutonium for military purposes. This component of the Calder Hall mission was deleted in 1995, when the United Kingdom ceased nuclear weapons production.

With the success of Calder Hall and *Nautilus* in the United States, the British government decided to design and build its own submarine reactor. The result of the effort was too big to fit in a naval vessel, but the British advanced gas-cooled reactor would become another successful public utility power source. The gas-cooled reactor, called "the golf ball" for its round shape, was a step forward in the sophistication of the economical

The Calder Hall Unit 1 at Sellafield, England. This was the first reactor to generate significant electrical power, but its graphite-moderated, gas-cooled design is now considered obsolete. (*Sellafield, Ltd.*)

graphite-moderated reactor designs preferred by the British. The coolant was carbon dioxide gas, blown through channels bored in the graphite pile, and it was a good design for safety. The carbon dioxide was nonflammable, and it could not flash explosively from liquid to gas, as could water.

With all the effort to keep down the cost of the nuclear-generating plants, the British government found that it still cost 25 percent more to produce power by nuclear means than it cost to burn coal, even given the bonus of plutonium production. The government, seeing a larger picture, decided in 1960 to promote nuclear power as an alternative to coal production so that all the fortunes of the United Kingdom would not depend on a single power source. Having an alternate source of electricity on the power grid would give them bargaining power against the coal miners' unions, which had given them reason to be concerned, beginning with a general coal miners' strike in 1926. In the longer view, the amount of coal that can be economical extracted in Great Britain is fixed and will not last forever. Nuclear fuel has a much longer life span.

The British nuclear industry built 11 power plants using variations and improvements of the original Calder Hall Magnox reactor. Two *Magnox* reactors were exported, one to Latina, Italy, and one to Tokai Mura, Japan. Nine reactors were built in France looking suspiciously like British Magnox designs, and three were built surreptitiously in North Korea using the declassified Calder Hall Magnox blueprints. All of these plants are now shut down. There are now seven nuclear plants operating in the United Kingdom. Six are advanced gas-cooled reactors, based loosely on the Magnox design, but now using enriched uranium oxide fuel. One is a standard Westinghouse pressurized water reactor, as pioneered by the *Nautilus* submarine program.

In a parallel program of nuclear engineering independence, Canada developed its own unique form of nuclear power plant. Great Britain avoided the high cost of building and running uranium enrichment facilities by purposefully designing reactors with high-efficiency graphite moderation. Canada wished also to avoid the cost of enrichment but chose heavy water as the high-efficiency moderator. This was the strategy that Germany had hoped to use during World War II, but Canada in the late 1950s assembled a partnership among Atomic Energy of Canada Limited and several private industries. They pooled resources and developed the *CANDU,* or CANada Deuterium Uranium, power plant.

Unlike the United States, Canada lacked the heavy industry necessary to build large steel pressure vessels that are used in pressurized and boiling water reactors. Instead, the heavy water moderator in a CANDU is contained in a low-pressure tank called the calandria, and the fuel is enclosed in small-diameter zirconium tubes. The tubes, which are easy to fabricate, conduct heavy water through the fuel at high temperature and pressure. The Canadian design is thus a pressurized water reactor that uses heavy water in the primary loop, through the fuel, and ordinary or light water through the secondary loop, making steam for the power-turbines. Refueling, which must be frequent due to the low U-235 content, is accomplished by automatic machines, pushing new fuel through one end of the reactor and catching it as it falls out the other.

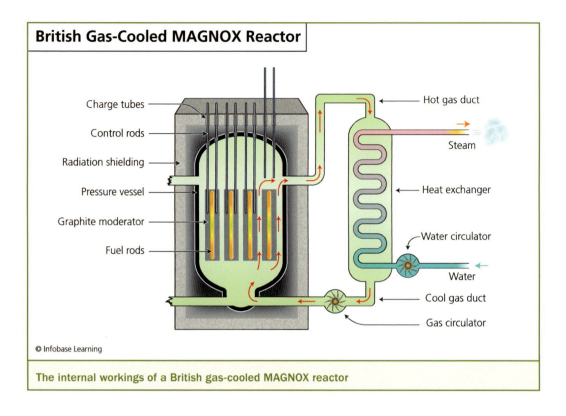

British Gas-Cooled MAGNOX Reactor

Charge tubes

Control rods

Radiation shielding

Pressure vessel

Graphite moderator

Fuel rods

Hot gas duct

Steam

Heat exchanger

Water circulator

Water

Cool gas duct

Gas circulator

© Infobase Learning

The internal workings of a British gas-cooled MAGNOX reactor

The first CANDU was built in 1962 in Rolphton, Ontario, and it ran for 25 years at low power, proving the concept of a heavy water power reactor. A second CANDU was built at Douglas Point in 1968, and further expansion in Canada and foreign sales have put 29 CANDU reactors in operation. There are 17 in Canada, and there are also CANDU reactors operating in South Korea, China, India, Argentina, Romania, and Pakistan. India has built 13 CANDU-derivative power plants without Canadian assistance.

A disadvantage to the CANDU design is the high cost of construction and building materials. Heavy water of sufficient purity costs about $1 per gram, and several metric tons are required in one reactor. As is the case with most nuclear reactor designs, the capital cost of building the power plant is 65 percent of the lifetime cost of producing power, with the cost of the fuel being less than 10 percent.

Other reactor designs have been tried with less success. Experimental reactors using plutonium-breeder technology or liquid-metal coolants, while having interesting potential for an economy in which uranium is not available, have proven less than practical for commercial power production. Some designs, such as the infamous Soviet RBMK graphite pile, have a lack of inherent safety characteristic and are being phased out as the plants reach the end of operating life. In general, the pressurized water reactors (PWRs) have proven to be the preferred design for practical power sourcing, with the boiling water reactor (BWR) a second choice. Japan, with 55 nuclear reactors in 17 power plants, is representative of the international commitment to alternative base power supplies. Of these 17 power plants, four were knocked out of commission by the Tōhoku earthquake of 2011. One, the Fukushima I Nuclear Power Station, was too damaged to be brought back into service.

The Qinshan Phase III Nuclear Power Units 1 and 2, in Zhejiang, China. These are CANDU-type reactors, supplied by Atomic Energy of Canada Limited. *(Atomic Energy of Canada, Ltd.)*

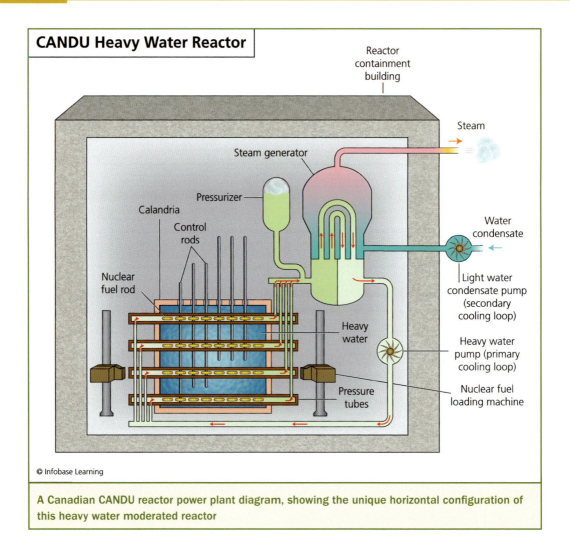

CANDU Heavy Water Reactor

Reactor containment building

Steam

Steam generator

Pressurizer

Calandria

Control rods

Nuclear fuel rod

Water condensate

Light water condensate pump (secondary cooling loop)

Heavy water

Heavy water pump (primary cooling loop)

Nuclear fuel loading machine

Pressure tubes

© Infobase Learning

A Canadian CANDU reactor power plant diagram, showing the unique horizontal configuration of this heavy water moderated reactor

France, where approximately 80 percent of the national electrical power is produced by nuclear fission, has 16 nuclear power plants, mostly PWRs. France's fast breeder reactor, the Superphoenix Nuclear Power Station, has the distinction of being the only commercial power plant to come under rocket attack by an eco-pacifist group. In 1982, five exploding warheads were fired into the reactor containment building using a Soviet-made rocket launcher. Credit for the attack was taken by the Swiss Green Party. The Superphoenix was shut down in 1997 because of nagging problems with the 6,063 tons (5,500 metric tons) of liquid sodium coolant leaking onto the floor.

Germany has 14 nuclear power plants, Russia has 10, and the United States has 51. A power plant usually has more than one reactor, and in the

United States there are 104 reactors. Nuclear plants in the United States produce just under 20 percent of the total electricity used by consumers. The lone sodium-cooled fast breeder reactor in the United States

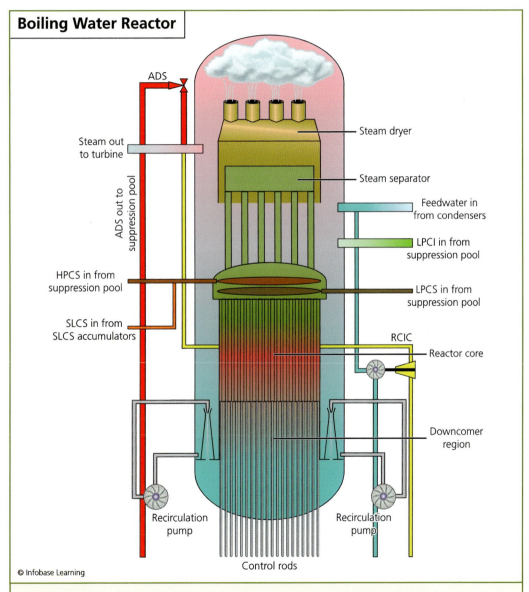

Boiling Water Reactor

ADS

Steam out to turbine

ADS out to suppression pool

Steam dryer

Steam separator

Feedwater in from condensers

LPCI in from suppression pool

HPCS in from suppression pool

LPCS in from suppression pool

SLCS in from SLCS accumulators

RCIC

Reactor core

Downcomer region

Recirculation pump

Recirculation pump

Control rods

© Infobase Learning

A boiling-water reactor diagram. Water comes in through the pipe on the right and leaves as steam through the pipe on the left. ADS, HPCS, SLCS, LPCI, LPCS, and RCIC are emergency core cooling systems. The recirculation pumps control the reactor by regulating the sizes of steam bubbles in the core.

commercial power grid, Fermi I near Monroe, Michigan, suffered a melt-down in 1966 and proved too expensive to be of practical use.

A single nuclear plant can produce as little continuous power as 137 megawatts by the KANUPP reactor in Pakistan or as much as 5,700 mega-watts by the Zaporizhzhia Nuclear Power Plant in the Ukraine. There are modest components of the national power supply dedicated to nuclear methods in Finland, Hungary, Brazil, Mexico, Bulgaria, Argentina, the Philippines, Romania, and South Africa. Spain has an impressive eight nuclear power plants, Sweden three, and Switzerland four.

THE ENVIRONMENTAL PROTECTION AGENCY AND LONG-TERM SPENT-FUEL STORAGE

A secondary engineering challenge for nuclear power production has been the safe handling and long-term storage of the dangerously radioactive products resulting from nuclear fission. The fission of a uranium nucleus results in two new lesser nuclei, of roughly half the weight of the original nucleus. These newly formed elements are always neutron heavy, having more neutrons than would normally be found in the nuclei. To reach a natural equilibrium, the new elements must decay radioactively, convert-ing excess neutrons to protons or occasionally ejecting delayed fission neu-trons. Powerful gamma rays result as the changing nuclear structures settle into the new equilibrium. The length of time for this process depends entirely on the species of new elements that are made by a given fission and vary over a wide range. The half-life, or time required for half of the radio-activity to decay away, can vary from microseconds to thousands of years. Having a short half-life means that a radioactive element, or radionuclide, will be extremely radioactive, but it will decay quickly. Having a long half-life means that a radionuclide will be slightly radioactive but long-lived.

The most radioactive waste products have decayed away six min-utes after fission. Iodine-131 and barium-140 are gone after four months. Cerium-141, zirconium-95, and strontium-90 take two or three years to disappear, and cerium-144, ruthenium-106, and promethium-147 linger for more than 10 years. *Strontium-90* and *cesium-137* are the most persis-tent and possibly dangerous fission products, each with a half-life of 30 years. Compared with the volume of waste produced by any combustion process, such as burning coal, the bulk of fission products is miniscule. If all the electrical power that a person consumes in a lifetime were pro-duced by nuclear fission, then the waste product from that production would fit in a Coke can and weigh 2.0 pounds (0.9 kilograms). If that electricity were produced by burning coal, the solid waste would be a small mountain weighing 68.5 tons (62.1 mt) and the gaseous carbon dioxide would weigh 77 tons (70 metric tons).

At the dawn of the nuclear era during World War II, hazardous nuclear waste from A-bomb production was dumped in the ocean, stored temporarily in liquid form in underground tanks, or simply allowed to dissipate into the atmosphere. Risky nuclear facilities for research, fuel or isotope production, or open-air tests were purposefully placed in low-habitation areas. When nuclear power became a privately owned, commercial product, the management of waste products had to become a priority, to be controlled and regulated by the federal government. In 1957, the National Academy of Sciences, after careful study and consideration, recommended that the best way to dispose of nuclear waste was to bury it in rock, deep underground.

In 1970, President Richard M. Nixon (1913–94) proposed the Environmental Protection Agency (EPA) to protect human health and safeguard the natural environment. Nixon signed this new agency into being on December 2, 1970, and an Office of Air and Radiation was opened. Under the subject of hazardous waste, the Nuclear Waste Repository Act, PL 97-425, was signed in 1982, taking responsibility for the long-term storage of nuclear power by-products.

President Gerald R. Ford (1913–2006) formally abolished the AEC in 1974. The AEC had been in charge of both promoting nuclear power and controlling nuclear power simultaneously, and this had long been seen as a conflict of interest. The commission was broken into two new agencies, the Energy Research and Development Agency (ERDA), for promotion, and the Nuclear Regulatory Commission *(NRC)*, for control. The NRC would write the rules and regulations for spent-fuel storage, and the ERDA would oversee research into the implementation of the tasks.

In 1977, in a proposal from President Jimmy Carter, the Department of Energy Organization Act, PL 95-91, dismantled ERDA and replaced it with the Department of Energy *(DOE)*. Radioactive waste disposal was clearly specified as a primary DOE responsibility. A Waste Isolation Pilot Plant (WIPP) had been in planning and design since 1974. After more than 20 years of scientific study, regulatory actions, and public debate, WIPP began operation on March 26, 1999.

WIPP is located 2,150 feet (655 m) underground, approximately 26 miles (42 km) east of Carlsbad, New Mexico. Radioactive waste is stored in rooms excavated out of the Salado and Castile Salt Formations. Storing waste in salt is considered ideal, because the formation has been stable and free of any moisture for 250 million years. Because there is no water

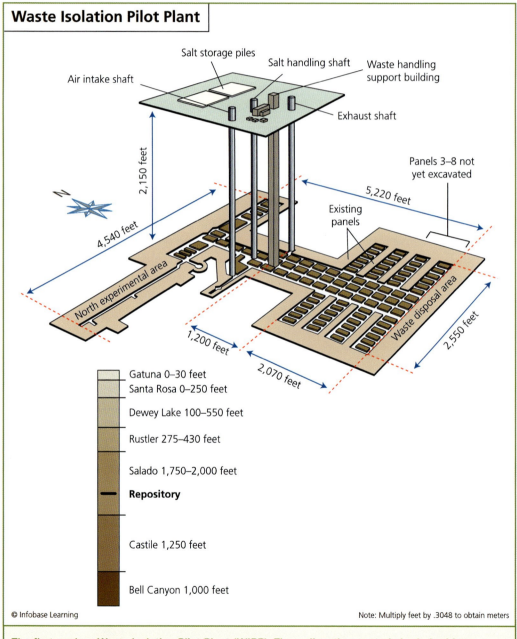

Waste Isolation Pilot Plant

Salt storage piles
Salt handling shaft
Waste handling support building
Air intake shaft
Exhaust shaft
2,150 feet
Panels 3–8 not yet excavated
5,220 feet
N
Existing panels
4,540 feet
North experimental area
Waste disposal area
2,550 feet
1,200 feet
2,070 feet

Gatuna 0–30 feet
Santa Rosa 0–250 feet
Dewey Lake 100–550 feet
Rustler 275–430 feet
Salado 1,750–2,000 feet
Repository
Castile 1,250 feet
Bell Canyon 1,000 feet

© Infobase Learning

Note: Multiply feet by .3048 to obtain meters

The first nuclear Waste Isolation Pilot Plant (WIPP). The radioactive waste is buried midway between the top of the Salado salt formation and the bottom of the Castile salt formation.

in a salt formation, nothing will dissolve and leak into the drinking water. If any cracks develop in a room made of salt, the plastic characteristic of the material will cause it to flow and fill any gap.

Although it is planned to hold only the radioactive waste from nuclear weapons production and not spent fuel from power reactors, the pilot plant is considered an excellent test of engineering and construction concepts for long-term storage. The facility is expected to continue accepting waste canisters until 2070. After it is sealed, radioactive monitoring will continue for another 100 years.

In 2028, a final plan will be submitted to the DOE for marking the site as a warning to future explorations. Warnings in seven languages will be etched into the floor of a large granite room over the portal. Pictographs showing a person screaming are also considered.

A permanent storage facility for spent reactor fuel was in development in the 1970s, but there is more to the task of radioactive waste. There must be a way to transport the spent fuel from power plants located all over the country to a central repository. Special, crash- and fireproof shipping casks were developed by the DOE, tested, and judged against 10 CFR 71 and the International Atomic Energy Agency standards, Regulations for the Safe Transport of Radioactive Material. The requirements are strict. A shipping cask must be strong enough to remain intact after one-hour immersion under 655 feet (200 m) of water, a 30-minute fire at 1,475°F (800°C), and a 30-foot (9-m) drop onto an unyielding surface. The NRC further requires that shipments follow only approved routes, have armed escorts, and notify in advance the states through which a fuel shipment will pass.

For two years, the DOE subjected spent fuel shipping casks to every form of train wreck, truck accident, and hostile action imaginable, and none of these tests resulted in a leaking container. More than 3,000 shipments of spent reactor fuel have been safely transported in the United States, as reactors have been shut down and dismantled. Canada has similar safety requirements for safe spent fuel shipping, as does the United Kingdom. British Nuclear Fuels Limited has transported more than 14,000 casks of spent fuel over rail, road, and water for Great Britain, Japan, and continental Europe.

With the transportation problem solved, Congress established a national policy to build an underground storage facility for spent fuel in the Nuclear Waste Policy Act in 1982. *Yucca Mountain,* a ridgeline in Nevada 80 miles (129 km) northwest of Las Vegas, was studied as an ideal location for deep storage since 1978. It is in a federally owned desert with no habitation. Starting with a thorough investigation of the area's geology, the DOE began design work for the spent fuel facility. On July 23, 2002, President George W. Bush (1946–   ) signed House Joint Resolution 87, allowing the DOE to apply for a construction license with the NRC.

In its final design, the Yucca Mountain Nuclear Waste Repository will hold 300 million pounds (136 million kg) of spent reactor fuel. The project has cost $9 billion, and the expense is borne by a tax on each kilowatt of power generated by nuclear means. The facility is scheduled to be completed in 2017, but its future is presently on hold. The state of Nevada has decided that nuclear waste from all over the country should not be buried there. Other countries, such as Japan, France, and the Netherlands, have also established their own long-term spent-fuel storage strategies, but the Yucca Mountain Repository is probably the largest and the most controversial in the world.

NUCLEAR POWER GOES INTO A LONG SLEEP

At 4:00 A.M. on March 28, 1979, a series of unfortunate events caused the core of reactor number 2 at the *Three Mile Island* Nuclear Power Station near Harrisburg, Pennsylvania, to melt. Although there were no casualties, the power unit was a total loss, and the psychological shock to the people in Pennsylvania and the rest of the country counteracted decades of nuclear power promotion.

For all the negative feelings generated by this accident in Pennsylvania, it was not the incident at TMI-2 that stopped the development of nuclear power in its tracks. Economic realities and the fact that the United States had accumulated an overcapacity of power generation had halted nuclear power in its tracks years before. Between 1973 and 1979, 40 nuclear power plant projects had already been cancelled. Since 1978, a year before the Three Mile Island accident, no new nuclear power plants had been authorized for construction in the United States. Of the 129 nuclear plants that had been cleared for building, only 53 projects were completed.

Starting with the Shippingport reactor in the late 1950s, nuclear power had seemed economically advantageous over coal-fired power production. Increasing regulatory issues, the rising cost of building sites, and the price of insurance had expanded the cost to the point where nuclear was no longer a bargain. The nuclear power building boom of the 1960s, anticipating a steadily increasing electrical load, had introduced too much generating capability into the economy. The effects of the TMI-2 accident affected the public mood concerning nuclear power. Before the incident, 70 percent of the general public in the United States favored nuclear power. Afterward, the support fell to 50 percent.

On April 27, 1986, a major incident unfolded that would affect the international mood toward nuclear power. A worker at the Forsmark Nuclear Power Plant in Forsmark, Sweden, showed up for work and walked through the radiation-detection portal into the plant. All nuclear workers are checked for radiation as they enter and leave a plant, to keep track of any radiation they may have picked up while working. The alarms went off. A quick check with handheld radiation detectors showed that he was tracking in radiation on his clothes.

This radiation alarm had nothing to do with nuclear activity at Forsmark. It was due to a major reactor explosion in the Soviet Union, near a town named Pripyat in the Ukraine, 680 miles (1,100 km) away. The day before, at 1:23 A.M. local time, the Chernobyl RBMK reactor no. 4 suffered a major power excursion, or uncontrollable increase, during a test of a safety system. An immediate steam explosion tore off the 2,205-ton (2,000–mt) reactor top and the roof of the building it was in and scattered radioactive fission products far and wide. Dust went thousands of feet straight up, into the upper atmosphere. When air hit the hot graphite moderator the remains of the reactor caught fire. There were 1,874 tons (1,700 mt) of graphite in reactor no. 4, and it took days for it to burn. Subsequent hydrogen explosions only added to the conflagration. Dust from the fire came down in Sweden a day later.

Of the 6,600,000 people living nearby the plant, 56 were killed directly by the accident, and an estimated 9,000 more contracted cancer from exposure to the massive radiation release. The city of Pripyat was quickly and permanently evacuated. Soviet engineering had designed the worst nuclear power reactor in the world, resulting in a radiation release that was so big it was difficult to measure. It made the next-worst disaster, the Windscale fire in England, seem small in comparison. The lasting result was a global slowdown in power plant expansion and a negative change in the collective opinion toward nuclear energy.

NEW REALITIES

The Chernobyl disaster may have marked the end of the long period of nuclear exuberance, in which experimental reactors were assembled in the desert, bolted into boats, shot into orbit, buried in arctic ice, and built to test exotic ideas. Every nuclear power reactor that was built in the United States was experimental. There was no standard design, and no two reactors were exactly alike. Even in a plant with two, identical-looking reactors, built by the same manufacturer, the two units were not exactly the same, requiring unique operator training and different spare parts. It was a massive engineering experiment in which much was learned, by trial, error, and accident investigation.

A point that was definitely learned was how not to build a nuclear reactor. So many designs that now seem obviously flawed had to be tried, and from all the large and small failures came a condensation of wisdom, pointing toward inherently safe, accident-free nuclear reactor designs. The path to a safe reactor is not necessarily coincident with a path to the least expensive reactor, and that is the hardest lesson of all to learn. The cheapest, simplest construction may also be the best construction, although not necessarily, and in nuclear power the path of success must follow the safety path.

Nuclear power has been in a holding mode for the past 30 years, neither moving forward nor receding. As it has remained dormant, the needs of the industrial world have changed. Atmospheric chemistry and general environmental pollution have become factors in power generation decisions. Global warming and air pollution are larger issues now than they were 30 years ago, and burnable materials such as coal, oil, and natural gas are now seen as finite resources. Because of all these factors, nuclear power generation is being given a second look.

On March 11, 2011, the expansion of nuclear power was once again challenged when the biggest earthquake in the history of Japan occurred off the northeast coast of the main island, Honshu. Minutes later, a powerful tsunami struck, wiping out entire towns and killing tens of thousands of people. Although designed for the maximum expected seismic activity, nuclear plants located on the beach faced the full force of the quake and tsunami. Closest to the quake, Onagawa, Tokai, and Fukushima II suffered complete automatic shutdowns. Fukushima I also shut down three reactors that were running but lost the use of its emergency diesel generators in the tsunami. There was no power to run coolant pumps, and the situation slowly progressed over days from serious to destructive. Reactor cores, still hot from having run continuously for years at full power, melted, and reactor buildings exploded from hydrogen gas buildups. Fission products escaped and polluted Japan and the Pacific Ocean. Repercussions and rebuilding from this massive human and economic disaster will continue for years.

Japan is recovering, and the need for nuclear power in this island nation and the world are still there. Expansion of nuclear power proceeds, using better and stronger reactors, designed for safe cooldown even when all power, including emergency generators, has been lost. Over the past 30 years, the nuclear industry has been refining and implementing safety procedures and systems, improving training, and always learning from mistakes and acts of nature, small, large, and calamitous.

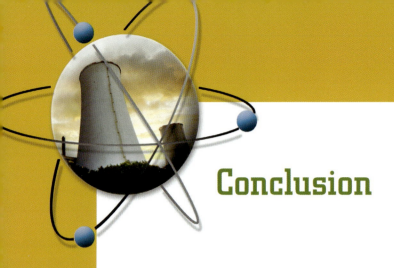

Conclusion

For hundreds of thousands of years, mankind has released stored energy by burning hydrocarbons, such as wood, natural gas, and oil. Heat is produced, as are chemical compounds such as water and carbon dioxide. This is a simple chemical process, employing the weak forces that hold together the electron structures of atoms. In the case of all energy conversion, there is direct matter-to-energy conversion. The products of combustion weigh less than the original components, but the effect is so slight it is not measurable. A much more obvious matter-to-energy conversion employs the forces that bind together the atomic nucleus and not just the atom. The nuclear forces are a million times greater than the electron forces, and the energy release is accordingly larger on an atom-by-atom basis.

Anything that will burn in air will give a combustion energy release, but to get a nuclear energy release is not so simple. The Sun and stars release energy by nuclear conversion, but the stellar process is difficult to scale down. One way that nuclear energy can be released with practical effect is to use fissile materials. Only a few special species, or isotopes, of a few elements are fissile, meaning that they have nuclei that will blow in half upon capture of a passing neutron, and further that the process of fission releases multiple neutrons. The neutron is a special, subnuclear particle, and its presence can affect nuclear properties. The products of a nuclear fission definitely weigh less than the original components, and the effect is measurable. Uranium-235, a rare isotope of uranium, is fissile.

The fission process releases energy at a very efficient level, more than a million times greater per atom than the best chemical process, and it is sustainable. Each fission is initiated by the capture of a free neutron, and each fission releases more than two new neutrons. These neutrons can then cause further fissions in other U-235 nuclei, and there are neutrons left over to waste. If there were only one neutron released for every neutron captured, then nuclear fission would never sustain. Out of trillions of fissions per second, if only one neutron were lost then the process would become nonsustaining and shut down. The existence of excess fission neutrons allows some to be lost by nonproductive capture and leakage from the assembly of uranium.

Energy release by fission uses a plentiful fuel, uranium. Using advanced fuel-breeding technology, there is enough uranium in the Earth's crust to supply the energy needs of mankind for thousands of years. An important advantage of nuclear energy release is that it results in no greenhouse gases, and the volume of the waste product is millions of times less than the waste product of any technology that involves combustion. An important disadvantage to nuclear energy release is that the waste products are dangerously radioactive. The elements produced by the splitting of a uranium nucleus are unnaturally neutron heavy. They tend to revert to a more natural state, and in doing so they release heat and *ionizing radiation*. These materials must be handled with unusual care and attention.

The special handling of waste and the extraordinary detail with which every aspect of this new energy conversion process must be conducted have taken time to work into the industrial culture. Becoming accustomed to the cost of such a fastidious process has been a challenge, but this is a new century with modified expectations, requirements, and anticipations. As mankind and civilization have matured, so has the concept of energy conversion and its relationship with the biosphere.

Chronology

550 B.C.E.   The earliest known mention of atomic structure in the Nyaya and Vaisheshika schools in ancient India is formulated, theorizing that elements combine into more complex objects, first in pairs, then in trios of pairs.

450 B.C.E.   The views of Leucippus are systemized by his student Democritus and recorded in ancient Greece; the word *atomos,* meaning "uncuttable," is first used.

1661   Robert Boyle publishes *The Sceptical Chymist,* in which he argues that matter is composed of various combinations of different "corpuscles" or atoms.

1789   Antoine Lavoisier coins the term *element,* meaning the basic substances that cannot be further broken down by chemistry.

1803   John Dalton uses the concept of atoms to explain why chemicals seem always to react in simple proportions and why certain gases dissolve in water more easily than others, explaining the concept of chemical compounds.

1827   The British botanist Robert Brown discovers Brownian motion by observing dust particles floating in water under a microscope.

1885   Johann Balmer, from Switzerland, predicts the spectral lines of hydrogen.

1887   Heinrich Hertz discovers "electric waves," or radio.

1890   Johannes Rydberg publishes a general formula for line spectra, using the Rydberg constant.

1895   On November 8, Wilhelm Roentgen discovers X-rays radiating from the fluorescing glass walls of a cathode-ray tube.

1896   Henri Becquerel discovers radioactivity on January 24.

1897   J. J. Thomson discovers the electron through work on cathode-ray tubes, proving that atoms can be broken down further.

1898   Marie and Pierre Curie purify radium from uranium ore.

1899	Ernest Rutherford names the alpha and the beta rays.
1900	Ernest Rutherford discovers a radioactive gas emanating from thorium.
	On October 19, Max Planck derives h, the "elementary quantum of action," or Planck's constant.
1903	Ernest Rutherford and Frederick Soddy publish a paper with the first calculations of the amount of energy released by radioactive decay, and Rutherford speculates that a wave of atomic disintegration might be started through matter, which would make the world go up in smoke.
	Hantaro Nagaoka postulates a saturnian model of the atom, with electrons orbiting in a flat ring around a positively charged particle.
1905	Otto Hahn comes to Montreal, Canada, to work with Ernest Rutherford.
1907	Hans Geiger and Ernest Marsden, working for Rutherford, discover alpha particles scattering more than 90 degrees in gold foil.
1908	Ernest Rutherford proves that an alpha particle is a charged helium nucleus.
1909	Ernest Rutherford discovers the nucleus, with the gold foil experiment, and proposes the planetary model of the atom.
1911	On March 7, Ernest Rutherford announces the discovery of the atomic nucleus at the Manchester Literary and Philosophical Society.
1913	Niels Bohr imposes quantum mechanics on the planetary model of the atom, restricting electrons to specific orbits.
1921	Construction is completed on January 18 for Niels Bohr's Institute for Theoretical Physics in Copenhagen.
1931	On December 28, Irène Joliot-Curie reports the penetrating effects of "beryllium radiation" (neutrons) to the French Academy of Sciences.
1932	James Chadwick completes his paper for *Nature,* "Possible Existence of a Neutron," in the February 17 edition.
1933	On September 12, Leó Szilárd thinks of the self-sustaining chain reaction while crossing Russell Square on Southhampton Row.
1934	In the spring, Leó Szilárd files a patent on the nuclear chain reaction.

1938   In December, Lise Meitner postulates nuclear fission, based on experiments of Otto Hahn in Germany.

1939   Lise Meitner and Otto Frisch publish the nuclear fission paper in January.

On March 17, Enrico Fermi meets with the chief of naval operations in Washington, D.C., to suggest nuclear weapons.

Paul Harteck writes to the German War Office in Berlin on April 24, telling of new developments in nuclear physics, making explosive devices possible.

On July 10, Leó Szilárd gets a rejection letter from the navy awarding no money for nuclear research.

On October 11, Leó Szilárd gains remote audience with President Franklin Roosevelt and presents the Einstein letter.

On October 21, Leó Szilárd, Edward Teller, and Eugene Wigner meet with the newly formed Uranium Committee and are awarded $6,000 to produce the atomic bomb.

1940   The $6,000 grant for atomic bomb research arrives on February 20.

On April 25, Albert Einstein posts another letter to FDR urging a more serious look at the danger from a German atomic bomb.

1941   Secret British findings are reported to James Conant and Vannevar Bush on July 10.

On December 7, Ernest Lawrence successfully separates U-235 from U-238 at Berkeley with the calutron.

1942   In summer, Igor Vasilevich Kurchatov is called to Moscow and assigned to build an atomic bomb for the Soviet Union.

On September 17, Leslie Groves is named head of the Manhattan Project (the S-1 Project) and is tasked with building a nuclear weapon.

On September 18, the U.S. government buys 1,250 tons of uranium ore from a Belgian mining company.

On November 16, Enrico Fermi and associates begin on Chicago Pile 1 (CP-1) at the University of Chicago.

On November 16, J. Robert Oppenheimer picks Los Alamos as the site for the bomb lab.

On December 2 at 3:53 P.M., CP-1, the world's first nuclear reactor, goes critical and generates half a watt of power.

On December 22, Yoshio Nishina starts the Japanese atomic bomb project at the Riken Institute in northwest Tokyo.

1943 The Los Alamos Lab opens for business on April 15, with three days of briefings at the main tech complex.

A U-235 separation plant at Oak Ridge, using Ernest Lawrence's magnetic apparatus developed at Berkeley, begins operation in August.

1944 On July 11, J. Robert Oppenheimer advises James Conant that the simple gun-assembly they had been counting on will not work with plutonium as the fissile material.

1945 On August 8, "Little Boy" is dropped on Hiroshima.

On August 11, "Fat Man" is dropped on Nagasaki. Both cities are destroyed.

1946 On August 1, President Harry S. Truman signs the Atomic Energy Act, placing the nuclear energy industry under civilian control.

1947 In October, the Atomic Energy Commission begins work on investigations into peaceful uses of nuclear energy.

1953 The first boiling water reactor experiment, BORAX-I in Idaho, proves that steam formation limits power and prevents runaways.

On December 8, the Atoms for Peace program is unveiled by President Dwight D. Eisenhower.

1954 In August, declassified government documents are made available to civilian nuclear energy programs by a major amendment to the Atomic Energy Act of 1946.

1955 In January, the Atomic Energy Commission begins cooperation with civilian industry to develop nuclear power.

On July 17, BORAX-III provides Arco, Idaho, with power for a population of 1,200 for one hour.

On January 17, the USS *Nautilus* SSN 571, the world's first nuclear-powered submarine, begins sea-trials with its Westinghouse pressurized water reactor, built at the Bettis Atomic Power Laboratory.

1957 The International Atomic Energy Agency is formed with 18 member countries to promote peaceful uses of atomic energy.

In September, President Dwight D. Eisenhower signs the Price-Anderson Act, protecting citizens, utilities, and contractors from lawsuits due to nuclear plant accidents.

On October 10, the Windscale graphite PU-production reactor in Great Britain catches fire, contaminating northern Europe with 20,000 curies of I-131.

On November 24, the first civilian power reactor, the Sodium Reactor Experiment, goes online in Santa Susana, California.

On December 2, the first full-scale nuclear power plant in the United States (60 MW) goes online at Shippingport, Pennsylvania.

1962   The first advanced gas-cooled reactor is built at Calder Hall, Great Britain, to power a naval vessel, for which it is found to be too big, and the reactor becomes a public utility for electrical power.

1966   On October 5, the Enrico Fermi fast breeder reactor experiences a core meltdown. The reactor is a total loss.

1970   In December, the U.S. Environmental Protection Agency (EPA) is formed, under President Richard M. Nixon.

1973   At a meeting in October, the Organization of Petroleum Exporting Countries (OPEC) cuts oil production by 25 percent.

1974   In March, the Atomic Energy Commission (AEC) establishes the Formerly Utilized Sites Remedial Action Program (FUSRAP) and offers cleanup to former Manhattan Project sites.

In October, President Gerald Ford abolishes the AEC and replaces it with the Energy Research and Development Administration (ERDA) and the Nuclear Regulatory Commission (NRC).

1976   In October, the Resource Conservation and Recovery Act (RCRA) is passed to protect the public from waste disposal operations.

1977   In April, President Jimmy Carter bans nuclear fuel reprocessing.

1978   On December 30, the Three Mile Island Unit 2 nuclear reactor near Harrisburg, Pennsylvania, begins commercial power production.

In November, the Uranium Mill Tailings Radiation Control Act directs the U.S. Department of Energy (DOE) to clean up the old uranium mines.

1979	On March 28, at 4:00 A.M., in one of two reactor units at the Three Mile Island Nuclear Plant in Harrisburg, Pennsylvania, a pump in the condensate polishing system stops running. The situation deteriorates, and the reactor loses water cover of the core and melts, leading to the worst nuclear disaster in U.S. history.
	In October after the meltdown, in reaction to public and political concerns over the TMI accident, the NRC creates the Institute of Nuclear Power Operations (INPO), to address safety and performance issues.
1980	In October, the DOE is directed to build a nuclear-waste solidification demonstration at the West Valley nuclear fuel reprocessing plant in New York.
	In November, the Hanford Plant in Washington State changes to double-shell waste storage tanks from single-shell tanks.
1981	President Ronald Reagan lifts the ban on reprocessing nuclear fuel in October, but no private funding steps up to the problem of spent fuel reprocessing.
1982	At Shippingport, Pennsylvania, the first commercial nuclear plant is shut down for decommissioning by the DOE.
1983	In January, President Ronald Reagan signs the Nuclear Waste Policy Act, establishing a timetable for designating underground nuclear waste storage facilities.
	In April, a report from the Atomic Industrial Forum finds no injuries or deaths resulting from the Three Mile Island meltdown in 1979.
1986	On April 26, an RBMK reactor at Chernobyl in the USSR runs away and explodes, sending massive contamination into the Northern Hemisphere.
	In May, Soviet medical experts predict an increase of approximately 30,000 cancer-related deaths over a 50-year period due to fallout from the Chernobyl disaster.
1987	In December, the Waste Policy Amendment Act designates Yucca Mountain, Nevada, for the first deep depository for high-level waste.
1989	Nuclear weapons production at the Rocky Flats Plant in Colorado and at the Fernald Feed Materials Production Facility in Ohio cease.
1992	In August, the first attempt at a uniform (standardized) nuclear plant design is submitted to the NRC.

1992   The Energy Policy Act is signed into law by President George H. W. Bush in October, reforming the licensing process for standardized nuclear plants.

In the same month, the Waste Isolation Pilot Plant (WIPP) Land Withdrawal Act withdraws public land for a nuclear waste repository in salt under the desert.

1994   In January, enriched U-235 is bought from Russia and down-blended for use in power plants, to keep it from being bought and used for bombs.

1996   In June, a class-action suit for damages from the Three Mile Island meltdown is dismissed by District Court Judge Sylvia Rambo, for scarcity of evidence.

On May 27, the last nuclear reactor built in the United States comes online and begins delivering power at the Watts Bar Nuclear Power Plant near Spring City, Tennessee. It took 22 years to build the plant.

2000   One hundred ten U.S. nuclear power plants achieve a world's record for reliability, having operated at 90 percent capacity for the past 10 years, generating 2,024.6 gigawatt-years of electrical power without incident.

2002   The U.S. House of Representatives approves Yucca Mountain, Nevada, as the final disposal site for all civilian nuclear power waste products.

In March, employees discover that acid has eaten through a reactor vessel cap at the Davis-Besse plant in Oak Harbor, Ohio. The owner, FirstEnergy Corp., pays a record $28 million fine.

2005   The only operating fast breeder reactor in the world, making more fuel than it burns, is the BN-600 in Beloyarsk, Russia. Japan, China, and India have plans for future breeder reactors.

2007   The NRC receives its first power plant construction application in the past 28 years, for two reactors near Bay City, Texas, by NRG Energy, Inc.

2009   In May, U.S. energy secretary Steven Chu announces that the Yucca Mountain Nuclear Waste Repository will no longer be considered as a disposal site for nuclear fission products, causing great concern and turmoil in the nuclear power industry.

In July, the House of Representatives voted 338 to 30 to include the funding for the Yucca Mountain repository in the FY2010 budget.

2011   March 11, Japan is hit with its biggest earthquake in history, followed quickly by a powerful tsunami. The Fukushima I nuclear power plant with six boiling water reactors experiences core meltdowns, hydrogen explosions, and is damaged beyond repair.

Glossary

activation making a substance artificially radioactive by bombarding it with neutrons

alpha particle also alpha rays, is a class of ionizing radiation composed of helium nuclei traveling at high speed. Alpha particles have a charge of +2 and are composed of two protons and two neutrons traveling stuck together. Alpha particles are highly energetic but have little ability to penetrate anything.

atom the smallest, most fundamental unit of an element, consisting of a central nucleus and a set group of orbiting electrons. The configuration of the electrons determines the chemical characteristics of the element.

atomic bomb or A-bomb, is an antiquated term meaning a nuclear weapon using a prompt, fast fission reaction in U-235 or Pu-239 as the explosive agent. A better term is nuclear weapon or nuclear device.

atomic energy an antiquated term meaning energy that is released by the fission of heavy nuclei or the fusion of light nuclei. The better term is nuclear energy.

beta ray or beta particle, is either an electron or a positron ejected from a decaying nucleus. If it is an electron, then a neutron has decayed into a proton. If it is a positron, then a proton has decayed into a neutron.

BORAX a series of experimental setups to test the concept of a boiling water reactor, near Arco, Idaho, at the National Reactor Test Station in the early 1950s

B-Reactor the first of three graphite moderated water cooled reactors built for the Manhattan Project near Richland, Washington, on the Columbia River to convert U-238 to Pu-239

breeder reactor a nuclear reactor that makes more fuel, through neutron capture in nonfissile nuclei, than it uses to produce power. Breeders typically transform U-238 into Pu-239, and Pu-239 can be used as fuel.

BWR or boiling water reactor, is a commercial power-production reactor in which light water is used as the moderator and the coolant. The moderator is allowed to boil in the reactor core, and the reactor uses no secondary coolant loop.

CANDU or CANada Deuterium Uranium, a class of heavy water moderated power reactors invented in Canada

cathode an electrode given a negative charge

cathode rays free electrons traveling through an evacuated tube

cathode ray tube a fully evacuated glass tube having an electrode embedded at each end. Applying a direct high-voltage current to the electrodes causes electrons, or cathode rays, to flow from the cathode to the anode terminals.

chain reaction a series of chemical or nuclear reactions in which each reaction causes another reaction

Chernobyl a city in the USSR, site of the worst nuclear disaster in history. In 1986, an RBMK reactor exploded, melted down, and caught fire, spreading radioactive debris over Europe.

cobalt-60 a radioactive isotope of cobalt. Cobalt-60, or Co-60, emits powerful gamma rays with a half-life of 5.27 years.

control rod a metal rod made of a neutron-absorbing metal, such as cadmium, used to soak up excess neutrons in a nuclear reactor and bring it to perfect criticality by adjustment

criticality the balance state of a nuclear chain reaction, in which the number of neutrons being lost through leakage, unproductive capture, or fissioning capture exactly equals the number of neutrons being produced by fission

critical mass the effective mass of uranium or plutonium fuel at which a nuclear reactor is critical. The mass is effective because it can be artificially adjusted using neutron-absorbing controls.

curie a unit of measure of radioactivity. One curie is 3.7×10^{10} radiation producing nuclear disintegrations per second.

cyclotron a device built to accelerate charged particles to high speed using an oscillating electrical field configured in a direction at a right angle to a static magnetic field. Cyclotrons were used to generate neutrons on demand and to activate U-238 into Pu-239.

deuterium heavy hydrogen. The deuterium is heavy because the nucleus contains both a neutron and a proton, and it weighs twice what an ordinary hydrogen nucleus weighs.

DOE Department of Energy

EBR an experimental breeder reactor. The EBR-I was built at Arco, Idaho, in 1950.

electromagnetic wave   an alternating electrical and magnetic field traveling through space at the speed of light

electron   a small, negatively charged particle of matter. Electrons occur in nature as components of atoms, and they are responsible for chemical interactions among atoms and the formation of chemical compounds.

enriched uranium   uranium reactor fuel that has had the U-235 content improved. A typical power reactor uses fuel with the U-235 artificially increased to 3 percent.

enrichment   the process of improving the concentration of fissile U-235 in natural uranium. Natural uranium, or uranium as mined, contains only 0.7 percent fissile U-235. The remainder is inert U-238.

eV   electron volt, or the amount of energy required to raise one electron to a potential of one volt

fissile   descriptor for an element that will release energy and excess neutrons when fissioned

fission   the splitting of a heavy nucleus into two lighter nuclei. Fission is caused by the absorption of a neutron in fissionable elements and can result in the release of excess energy.

fissionable   a descriptor for an element that can be fissioned by neutron capture

fission products   the lighter, always radioactive isotopes into which a fissile fuel breaks upon fission. Fission products are a wide range of isotopes, with half-lives from a few seconds to a few thousand years.

fluorescence   an energy conversion event at the atomic level, causing a visible glow. Ultraviolet light, for example, will make fluorescent dye glow, as the energetic ultraviolet photon is down-converted to a visible photon, re-emitted by the dye.

fusion   the combining of two light nuclei into one heavier nucleus, resulting in a release of excess energy

gamma ray   a high-energy electromagnetic wave, above X-rays on the electromagnetic energy spectrum, originating in the nucleus. A gamma ray is generated when the nucleus experiences a rearrangement of subnuclear particles.

Geiger counter   or Geiger-Mueller counter, is an electronic radiation detector used to measure the presence of gamma or beta rays. The Geiger counter makes use of the extreme amplifying properties of an avalanche effect in a gas-filled tube excited by a high voltage.

half-life the time required for a radioactive sample to decrease its level of radioactivity by one half

heavy water deuterium oxide, or water made with two deuterium atoms and one oxygen in each molecule

hydrogen bomb or H-bomb, is a nuclear weapon using the fusion of deuterium and tritium as the explosive agent. The fusion event requires an atomic bomb as an initiator to establish the necessary temperature and pressure.

implosion the initiation of explosive burning on the entire outside surface of a sphere of explosive material. A component of the resulting shock wave travels inward, toward the center of the sphere, gaining in power as the fixed energy of the wave is confined in a smaller and smaller volume.

ionizing radiation radiation of sufficient power to knock the top electron out of an atom upon collision. Examples of ionizing radiation are gamma rays, beta rays, and alpha rays.

isotope a subspecies of an element, distinguished by the number of neutrons in the nucleus. All possible isotopes of hydrogen, for example, have zero, one, or two neutrons in the nucleus, all of which have only one proton.

light water ordinary water, as is available from a municipal tap. Light water contains only traces of heavy water, or deuterium oxide.

Los Alamos National Laboratories a research and development facility set up during World War II for the atomic bomb project, on a desert mesa near Los Alamos, New Mexico

Magnox a now obsolete British reactor fuel formula, using magnesium clad uranium

Manhattan Project or Manhattan Engineering District, the secret atomic bomb development project of the U.S. Army Corps of Engineers during World War II

meltdown the damage that occurs to the core in a nuclear reactor if the temperature exceeds the melting point of the fuel, the fuel cladding, or the fuel support structures

MeV million electron volts, as a unit of energy applied to subatomic or subnuclear particles in motion

moderator   any substance used in a nuclear reactor to slow high-speed neutrons from fission down to thermal speed. Neutrons at thermal speeds are more likely to initiate fission in a reactor.

Nautilus   the first nuclear-powered object to move under its own power. *Nautilus* was a nuclear-powered navy submarine carrying conventional warhead torpedoes, launched in 1954.

neutron   a particle of matter having no electrical charge. Neutrons are components of the nucleus in an atom.

NRC   Nuclear Regulatory Commission, charged with the regulation and oversight of all nuclear activities in the United States

NRX   or National Reactor X-perimental, an experimental heavy water moderated reactor built in Canada at Chalk River in 1947

nuclear physics   the study of forces, objects, and fields involved in the atomic nucleus and its interactions with matter

nucleus   (pl. nuclei) the massive center of an atom, built of protons and in all but one case, neutrons. Hydrogen is the only atomic nucleus having no neutrons and only one proton.

Oak Ridge National Laboratories   a research, development, and production facility set up in a valley near Oak Ridge, Tennessee, during World War II for the atomic bomb project. Oak Ridge was headquarters of the Manhattan Project and site of the first uranium-enrichment facilities.

photon   a quantum of electromagnetic radiation. Albert Einstein referred to a photon as a "packet of energy."

pile   an archaic term, meaning nuclear reactor. The first reactors were literally piles of graphite bricks.

plutonium   element number 94 in the table of the elements. Plutonium is a chemically poisonous metal and is exceedingly rare in nature. Plutonium is commonly made by activating U-238 into neptunium, which then decays to plutonium.

proton   a particle of matter having a positive charge. Protons are components of the nucleus in an atom.

Pu-239   a fissile isotope of plutonium. One Pu-239 nucleus contains 94 protons and 145 neutrons.

Pu-240   a fissile isotope of plutonium. One Pu-240 nucleus contains 94 protons and 146 neutrons. Pu-240 is sensitive to spontaneous fission, not necessarily requiring a neutron capture to initiate the reaction.

PWR   or pressurized water reactor, is a form of commercial power reactor using ordinary water as the moderator, held at high pressure so that it remains in the liquid state

quantum   the smallest possible size of anything. A molecule, for example, is the smallest possible quantum of water.

quantum mechanics   the study of forces, objects, and fields at the lower limit of size, or the quantum level. Mechanics at and beyond the quantum level are remarkably different than at the continuum level, at which humankind normally interacts with matter.

radiation   a class of energy transmission by electromagnetic waves or by direct particle transfer

radioactive   capable of emitting radiation at a predictable rate by nuclear decay

radioactive decay   the tendency of certain isotopes to undergo change in the nucleus. Any change in the nuclear structure causes radiation to be emitted from the nucleus. The time at which the change occurs is completely random and unpredictable, yet the rate at which a large sample of the particular isotope will decay is predictable and is characteristic of the isotope.

radioactivity   the emission of radiation, either by the willful manipulation of a nucleus or by spontaneous nuclear decay

RBMK   a now obsolete power reactor design in the former Soviet Union. The RBMK is a graphite moderated boiling water reactor.

reactor   a machine or system built to sustain a neutron chain reaction in a fissile material for the purpose of power production

Santa Susana Field Laboratory   A rocket and nuclear reactor test facility in Moorpark, California. Santa Susana was the site of the sodium reactor experiment, the first commercial nuclear power reactor in the United States.

Sellafield   the site of British nuclear power and weapons development, near the village of Seascale on the coast of the Irish Sea in Cumbria, England. Sellafield was the site of Calder Hall, the first commercial nuclear power station.

thermal speed   the speed at which air molecules move at room temperature as they bounce around and hit each other. At cold temperatures, air molecules move more slowly and at very high temperatures more quickly.

Three Mile Island (TMI) an island in the Susquehanna River in Pennsylvania, near Harrisburg, and location of the Three Mile Island Nuclear Generating Station. A reactor at the Three Mile Island station experienced a meltdown in 1979.

Trinity the codename of the first nuclear weapons test. It took place in Alamogordo, New Mexico, on July 16, 1945.

U-235 a fissile isotope of uranium, having 92 protons and 143 neutrons in its nucleus. U-235 occurs rarely in nature, making up only 0.7 percent of the uranium found in the Earth's crust.

U-238 a non-fissile isotope of uranium, having 92 protons and 146 neutrons in its nucleus. U-238 makes up most of the uranium occurring in nature. It can decay indirectly into Pu-239 upon neutron capture.

ultraviolet light light that is beyond the frequency of violet light and is therefore invisible to human eyes

uranium element number 92 in the table of the elements. Uranium is common in nature and can be found in traces in most of the Earth's crust. It is mildly radioactive.

Windscale a pair of air-cooled graphite reactors built by the British government in 1946 to convert U-238 to Pu-239. A Windscale reactor caught fire in 1957.

X-ray an electromagnetic wave above light in the electromagnetic energy spectrum

Yucca Mountain or Yucca Mountain Repository, a deep geological repository storage facility for radioactive waste from power production located within a ridgeline on government-owned property in south-central Nevada

Further Resources

BOOKS

Albright, Joseph, and Marcia Kunstel. *Bombshell: The Secret Story of America's Unknown Atomic Spy Conspiracy.* New York: Times Books, 1997. Fresh information regarding a recently discovered Soviet spy in the World War II atomic bomb development project.

Allen, Thomas B., and Norman Polmar. *Rickover: Father of the Nuclear Navy.* Dulles, Va.: Quicksilver Books, 2007. A biography of Rickover, a complex individual who was responsible for successful civilian nuclear power as well as the nuclear-powered navy.

Blair, Clay. *The Atomic Submarine: The Story of the* "Nautilus," *the World's First Atomic Driven Vessel.* London: Odhams Press Limited, 1955. A rare account of the steps to develop the nuclear-powered submarine and its associated pressurized water reactor.

Cravens, Gwyneth. *Power to Save the World: The Truth about Nuclear Energy.* New York: Knopf, 2007. This book gives a current view of the state of nuclear power, considering all sides of the arguments for and against it as a practical source of electrical power.

Dewar, James A. *To the End of the Solar System: The Story of the Nuclear Rocket.* Ontario, Canada: Apogee Books, 2007. The history of nuclear rocket engine development, including descriptions of the technical and political hurdles.

Fuller, John G. *We Almost Lost Detroit.* New York: Ballantine Books, 1975. Clear descriptions of some of the most serious nuclear reactor near-disasters in the 1960s, including the Windscale fire and the SL-1 explosion in Idaho.

Glasstone, Samuel. *The Effects of Nuclear Weapons.* Washington, D.C.: U.S. Government Printing Office, 1957. The definitive study of how nuclear explosions in the atmosphere will affect people and objects exposed to the blast and subsequent radiation. Also available online. URL: http://www.princeton.edu/sgs/publications/articles/effects/.

Gleick, James. *Genius: The Life and Science of Richard Feynman.* New York: Pantheon Books, 1992. An extensive biography of Richard Feynman, an important member of the theoretical physics team at Los Alamos during World War II and a productive physicist and lecturer after the war.

Goodchild, Peter. *Edward Teller: The Real Dr. Strangelove.* Cambridge, Mass.: Harvard University Press, 2004. The most recent and complete biography of Edward Teller, an important scientist and controversial personality in the quest for nuclear domination of the world.

Goudsmit, Samuel A. *Alsos*. Woodbury, N.Y.: AIP Press, 1996. Goudsmit's personal memoir of his mission to Germany in World War II to uncover the German effort to develop an atomic bomb. The mission was code-named Alsos.

Groves, Leslie R. *Now It Can Be Told: The Story of the Manhattan Project*. New York: Harper, 1962. A history of the development of the atomic bomb by the head of the project, written after some details were declassified.

Isaacson, Walter. *Einstein: His Life and Universe*. New York: Simon and Schuster, 2007. An extensive biography of Albert Einstein with clear explanations of his pivotal theories.

Kelly, Cynthia C. *The Manhattan Project: The Birth of the Atomic Bomb in the Words of Its Creators, Eyewitnesses, and Historians*. New York: Black Dog & Leventhal, 2007. A collection of historical documents, memoirs, and personal accounts from people who worked on the Manhattan Project.

Mahaffey, James. *Atomic Awakening: A New Look at the History and Future of Nuclear Power*. New York: Pegasus, 2009. An expanded history of nuclear power, including recently declassified information from the nuclear weapons programs.

Moss, Norman. *Klaus Fuchs: The Man Who Stole the Atom Bomb*. New York: St. Martin's Press, 1987. A detailed account of the primary Soviet mole in the Manhattan Project.

Powers, Thomas. *Heisenberg's War: The Secret History of the German Bomb*. New York: Knopf, 1993. Details of the other side of the atomic bomb development during World War II, as it progressed and failed in Germany.

Reed, Thomas C., and Danny B. Stillman. *The Nuclear Express: A Political History of the Bomb and Its Proliferation*. Minneapolis, Minn.: Zenith Press, 2009. A disturbing chronology of the development and dispersal of nuclear weapons technology from World War II to the present, containing new and previously unpublished information.

Rhodes, Richard. *The Making of the Atomic Bomb*. New York: Simon and Schuster, 1986. This is a definitive history of the Manhattan Project, containing depth and details not available earlier.

Serber, Robert. *The Los Alamos Primer: The First Lectures on How to Build an Atomic Bomb*. Los Angeles: University of California Press, 1992. A reprint of the mimeographed notes given to physicists and chemists hired for top-secret work on the atomic bomb during World War II. Previously classified SECRET.

Smyth, Henry DeWolf. *Atomic Energy for Military Purposes: The Official Report on the Development of the Atomic Bomb under the Auspices of the United*

States Government, 1940–1945. Princeton, N.J.: Princeton University Press, 1945. The final report of the U.S. Army Corps of Engineers project to develop nuclear weapons. Also available online. URL: http://nuclearweaponarchive. org/Smyth/index.html.

Sparks, Ralph C. *Twilight Time: A Soldier's Role in the Manhattan Project of Los Alamos.* Los Alamos, N. Mex.: Los Alamos Historical Society, 2000. A personal account of what it was like to work under top-secret conditions during World War II on the Manhattan Project. This work is part of an ongoing effort by the Los Alamos Historical Society to document the important events during the war before human memories are lost.

Tucker, Todd. *Atomic America: How a Deadly Explosion and a Feared Admiral Changed the Course of Nuclear History.* New York: Free Press, 2009. An interesting placing of Hyman Rickover and the SL-1 reactor explosion in the center of the development of nuclear reactor safety procedures.

Wyden, Peter. *Day One: Before Hiroshima and After.* New York: Simon and Schuster, 1984. An additional history of the Manhattan Project. Each account of the atomic bomb development gives a slightly different approach and additional details.

Zoellner, Tom. *Uranium: War, Energy, and the Rock that Shaped the World.* New York: Penguin Books, 2009. A well-researched history of uranium, its uses, and its political power in the current state of world affairs.

WEB SITES

Further depth on many topics covered in this book has become available on the World Wide Web. Biographies of famous scientists are particularly available, and a deeper probing into the lives and the careers of notable physicists and chemists can be worth pursuing. Always look for cross-referencing links in a Web site. A double click on a highlighted word can take you even further, giving details on interesting subtopics. Also available are detailed Web sites for any government facility or agency, some giving access to archives and histories.

Antoine-Henri Becquerel. Available online. URL: http://web.lemoyne. edu/~giunta/becquerel.html. Accessed January 5, 2011.

This is a site containing full transcripts of English translations of papers read by Becquerel to the French Academy of Science in 1896. Nothing is better than reading the exact papers written by scientists to find details of historical experiments, and this site is valuable to those interested in knowing the events leading to the discovery of radiation.

AtomicBombMuseum.org. Available online. URL: http://atomicbomb museum.org/. Accessed November 16, 2010.

    This is a Japanese site providing an excellent record of the bombing of Hiroshima, Japan, January 5, 2011 at the close of World War II. Included are historical records, accounts of survivors, and a tour of the Hiroshima Atomic Bomb Museum, as well as discussions of problems faced by the children of atomic bomb survivors.

Atomic Energy for Military Purposes (The Smyth Report). Available online. URL: http://www.atomicarchive.com/Docs/SmythReport/index. shtml. Accessed January 5, 2011.

    This is an example of an out-of-print book that is currently available free on the Web. It is the complete historical account of the Manhattan Project by Henry DeWolf Smyth, as referenced in the Further Reading section above.

Code of Federal Regulations. Available online. URL: http://www.gpoaccess. gov/cfr/. Accessed January 5, 2011.

    The entire code is a vast set of documents, but with the search engine at this site it is possible to examine individual codes. Start with Title 10 CFR 50 to find what the Federal Government requires of anyone working with radioactive materials or nuclear technology. Instructions are provided for retrieving, browsing, searching, and linking to the codes, as well as the new e-CFR resource.

Department of Energy Hanford Site. Available online. URL: http://www. hanford.gov/. Accessed January 5, 2011.

    This is the official Web site of the Hanford Facility, which was the production point for plutonium during World War II. The Hanford Site is still in operation, and this complete Web site contains everything from online videos to an abbreviations and acronyms directory. Hanford's history also includes the Fast Flux Test Facility, an account of which is covered in some detail.

Farmhall. Available online. URL: http://www.atomicarchive.com/Docs/ Hiroshima/Farmhall.shtml. Accessed January 5, 2011.

    This is an excerpt from the declassified transcripts of secretly recorded conversations at Farm Hall. Ten captured German scientists were interned at Farm Hall, England, following the surrender of Germany in World War II, and these transcripts are very interesting, showing the extent of the German atomic bomb program.

LANL Research Library. Available online. URL: http://library.lanl.gov/. Accessed January 5, 2011.

This is a vast collection of e-books, databases, and news concerning all issues of nuclear science, nuclear technology, the history of nuclear topics, and current research and development at the Los Alamos National Laboratory. Students of nuclear science will find it an invaluable source of information. The library search can find books, journals, patents, reports, videos, audio tapes, and recommended Web sites.

Leó Szilárd Online. Available online. URL: http://www.dannen.com/szilard. html. Accessed January 5, 2011.

This is a comprehensive look at the world of the nuclear physicist, biophysicist, and "scientist of conscience" Leó Szilárd. Szilárd invented the concept of the nuclear chain reaction, as well as the cyclotron particle accelerator, the electron microscope, and a refrigerator in collaboration with Albert Einstein.

Los Alamos National Laboratory. Available online. URL: http://www.lanl. gov/. Accessed January 5, 2011.

This an online science and technology magazine, with excellent access and interest for the educator and students. It includes articles concerning current work at Los Alamos National Laboratory, with sections concerning the environment, business and technical transfers, and postdoctoral studies. A link to the Bradbury Science Museum is included in the features of this site.

Marie Curie. Available online. URL: http://www.aip.org/history/curie/. Accessed January 5, 2011.

This is a book-length biography of Marie Curie, with many pictures, rare photographs, and some technical and personal details of Curie's life as a scientist in France.

Nobel Prize. Available online. URL: http://nobelprize.org/nobel_prizes/. Accessed January 5, 2011.

This is the definitive site giving details of all recipients of the Nobel Prize in all categories, including biographies. It is the official Web site of the Nobel Prize Foundation. Many Nobel Prizes in physics and chemistry have been awarded for work in nuclear science.

Nuclear Weapon Archive. Available online. URL: http://nuclearweaponarchive.org/. Accessed January 5, 2011.

This a complete guide to nuclear weapons, including an archive of documents from the weapons developments and detailed accounts of the history of the atomic bomb. Also included are the histories of atomic bombs in other countries, including the Soviet Union, the United Kingdom, France, China, India, Pakistan, and North Korea. Current bomb inventories in the arsenals of declared nuclear states are listed, as well as undeclared nuclear states.

Oak Ridge National Laboratory. Available online. URL: http://www.ornl.gov/. Accessed January 5, 2011.

This the official site for the old Manhattan Project headquarters and the uranium enrichment plants from World War II. The laboratory is still in business, and this site tells all about it, with videos, photos, and accounts of ongoing nuclear and biological research.

Periodic Table of the Elements. Available online. URL: http://periodic.lanl.gov/default.htm. Accessed January 5, 2011.

This service is provided by the Los Alamos Laboratory, and it is a remarkably complete source of information concerning the elements of which all matter is made. A click on any of the 118 elements in the opening page will link to further information. It is particularly suited for elementary, middle school, and high school students.

Sellafield Ltd. Available online. URL: http://www.sellafieldsites.com/. Accessed January 5, 2011.

This site gives a good look into the involvement and attitudes of Europe in the worldwide nuclear industry. Although it covers only one nuclear facility in England, it is a good start in understanding the participation of Europe in the long-term energy solution of nuclear power. It includes the history of Sellafield, with a video of its construction, to the current uses of the facilities at Sellafield and the decommissioning of the Calder Hall reactors.

Timeline of the Nuclear Age. Available online. URL: http://www. atomicarchive.com/Timeline/Timeline.shtml. Accessed January 5, 2011.

This is a sequential listing of the important events in the development of nuclear technology, starting in the 1890s. There are many nuclear time lines available on the Web, but this one is particularly detailed, breaking the sequence into decades.

Tokyo Electric Power Company (TEPCO). Available online. URL: http:// www.tepco.co.jp/en/index-e.html. Accessed April 5, 2011.

The latest information concerning the earthquake destruction of the Fukushima I power plant is available at this site. It includes a free downloads library, frequent press releases, and up-to-date environmental radiation measurements.

Trinity Atomic Web Site. Available online. URL: http://www.abomb1.org/. Accessed January 5, 2011.

This an archive of atomic bomb documents. In the spirit of the Gutenberg Project, this site makes available all possible U.S. government documents concerning nuclear weapons.

United States Department of Energy. Available online. URL: http://www. doe.gov/. Accessed January 5, 2011.

This site spans the wide interests of this government agency, from energy science and technology to national security of energy sources. The Department of Energy owns, secures, and manages all the nuclear weapons in the military inventory. Although it concerns all sources of energy, nuclear power is a large component of the mission of the Department of Energy. This site is particularly accessible to students and educators.

United States Nuclear Power Federal Regulations, Codes, and Standards Users Group. Available online. URL: http://www.usnrc.org/. Accessed January 5, 2011.

This an interactive site that provides online technical forums and information on all federal regulations, codes, and standards for the nuclear power industry. The site includes searchable databases, quick links, and information resources on all nuclear issues.

United States Nuclear Regulatory Commission. Available online. URL: http://www.nrc.gov/. Accessed January 5, 2011.

This the most comprehensive government site for information concerning the control and safety regulation of the nuclear power industry. It provides a great depth of information concerning nuclear reactors, nuclear materials, radioactive waste, and nuclear security. There is a section that covers public meetings and involvement in nuclear power, as well as a constantly updated event report and news section.

W. C. Roentgen and the Discovery of X-rays. Available online. URL: http://www.medcyclopaedia.com/library/radiology/chapter01.aspx. Accessed January 5, 2011.

This a detailed account of Roentgen's life and his work discovering artificially induced radiation. General Electric, a major producer of X-ray equipment, has provided this information to Medcyclopaedia.

Index

Italic page numbers indicate illustrations.

A

activation 52–54

AEC *See* Atomic Energy Commission (AEC)

aircraft carriers 113–114

Air Force, U.S. 90–92, *91m*

Alamagordo test site 88, *89,* 90

alpha particles (alpha rays) 19–21, 23–26, *24, 37,* 135

Alsos Mission 83–84

AM-1 nuclear power station 101

Anderson, Herb 66

Argonne Laboratory 76

Argonne National Laboratory 105

atom (atomos), use of term 2–3, 5

atomic bombs
  chronology 136–138
  criticality 79–80
  Fermi's invention of the nuclear reactor 50–55
  first nuclear reactor (Chicago Pile 1) 64–70, *66, 67*
  German program 48–50, 83–85, *84*
  Japanese program 85
  Jewish refugees and 44, 50, 52–53, 55–56
  Manhattan Project *See* Manhattan Project
  neutron speed and interaction experiments 52–54
  nuclear power reactor v. 79–80
  as possible use for nuclear fission 45–46, 59–60
  production cessation 139
  secrecy 63–64, 81–82
  U.S. government involvement 60–64

Atomic Energy Act (1946) 137

Atomic Energy Commission (AEC)
  abolishment of 125
  chronology 137, 138
  Nautilus and 95, 99, 104–105
  Strauss as head of 109–111

Atomic Energy of Canada Limited 119

atomic nucleus, discovery of 18–26 *See also* quantum mechanics
  Lenard and nature of atoms 21–23
  Rutherford's alpha and beta rays (particles) 19–21
  Rutherford's discovery of (Gold Foil Experiment) 22, 23–26, *24,* 135

atomic structure, theories of 1–17
  astronomical model 26, *26,* 30–31, 135
  Becquerel rays (radioactivity) 10–12
  Bohr's model of electron orbit (quantum mechanics) 30–34, *31*
  Boyle's chemistry book 3–4, 134
  the Curies and radioactivity 15–17
  Dalton's principles of 4–5
  Faraday's experiments with electromagnets 5, *7*
  Hertz and radio waves 8
  Lavoisier's table of elements 4
  Leucippus' use of term *atomos* 2–3, 134
  Maxwell's equations 7–8
  neutrons, discovery of 35–41
  nucleus *See* atomic nucleus, discovery of
  at Nyaya and Vaisheshika schools 2, 134
  plum pudding model 14, 21, *26*
  Roentgen's X-rays 8–10, *9*
  Thomson's corpuscles (electrons) 12–15

atomic weight 35–36

atomos 2–3, 134
Atoms for Peace 105, 115, 137
atom smasher 57–59, *58*

B

Balmer, Johann 30–31, 134
Balmer Series 31
barium 43, 45
Becquerel, Antoine-Henri
    10–12, 15, 17, 134
beryllium 37–38, 135
beta decay 50, 51
beta particles (rays) 19–21, 135
Bethe, Hans 55, 78
big science 56
blackbody radiation 28–29
BN-600 fast breeder reactor
    140
Bohemian Grove (Bohemian
    Club) 70
Bohr, Niels 30–34, 52–53, 63,
    135
boiling water reactors (BSRs)
    107–108, 121, *123,* 137
BORAX reactors 106–108,
    107, 137
Boyle, Robert *3,* 3–4, 134
B-Reactor 76–78, 79
breeder reactors 106–108, *107,*
    138, 140
Briggs, Lyman 70
Brown, Robert 134
Brownian motion 134
BSRs (boiling water reactors)
    107–108, 121, *123,* 137
Bush, Vannevar 136

C

Calder Hall nuclear plant
    103, 116–117, *118,* 119, *120,* 138
California, University of,
    Berkeley 57–58, *58*

calutrons 58, 74–75, *75,* 136
Canada, nuclear power
    program 103–104, 119–120,
    122
CANDU 119–120, *122*
Carlsbad radioactive waste
    site 125–127
Carter, James Earl "Jimmy"
    104, 138
cathode rays/cathode ray
    tubes 9–10, 12–15, *14,* 21–23,
    134
cesium-137 124
Chadwick, James 36–38, *38,*
    135
chain reactions, self-
    sustaining *See also* atomic
    bombs; nuclear power
    plants/nuclear reactors
        criticality 6, 79–80
        prehistoric 6
        process of 66
        Szilárd, Leó 38–41, *40*
        weapon theory 59–60
Chernobyl accident 101, 129,
    139
Chicago Pile 1 (CP-1) 55,
    64–70, *66, 67,* 136
China, nuclear power 120, *121*
Chu, Steven 140
climate change 130
coal 115–116, 118–119, 125
cobalt 27–28
compounds 3–4, 5, 34, 134,
    144
Compton, Arthur 33, 65, 66,
    70
Conant, James 70, 136, 137
CP-1 (Chicago Pile 1) 55,
    64–70, *66, 67,* 136
criticality 6, 79–80
Curie, Marie and Pierre
    15–17, *16*
cyclotrons 57–59, *58, 78*

D

Dalton, John 4–5, 134
Daniels, Farrington 96
Daniels Pile 96
Davis-Besse plant 140
decay *See* radioactive decay
delayed criticality 79
Democritus 2–3, 134
Department of Energy, U.S.
    (DOE) 125, 127, 128, 138, 139
Desert Test Station (Atomic
    Energy Commission) 99
deuterium-oxide (heavy
    water) moderation 49, 104,
    119–120, *122*
Dick, Ray 95
DOE (Department of Energy,
    U.S.) 125, 127, 128, 138, 139
double-slit experiment
    32–33, 34
D-Reactor 78
Dunford, James H. 95
Duquesne Light Company
    114

E

E. I. du Pont de Nemours and
    Company 77
E = hv 29
Einstein, Albert 29, *29,*
    61–62, 64, 136
Eisenhower, Dwight D. 105,
    115, 137, 138
Electric Boat Company 99
electromagnetic waves 5, 7–8,
    18–19
electrons 9, 12–15, 30–32, *31*
electron volts 54
elements
    Dalton's atomic theory 4–5
    four basic (earth, air, fire,
        water) 2, 3
    periodic table 4, 33–34, 134

Energy Policy Act (1992)
140
Energy Research and
Development Agency
(ERDA) 125, 138
energy states 31, *31*
England *See* United
Kingdom
Enola Gay 90
Environmental Protection
Agency, U.S. (EPA) 124–
128, 138
ERDA (Energy Research and
Development Agency) 125,
138
Experimental Breeder
Reactor One (EBR-1) 106

F

Faraday, Michael 5, *7*
Fat Man (implosion design
atomic bomb) 80, 82–83, *87,*
87–88, 93, 137
Fermi, Enrico
at Argonne Laboratory
76
atomic bomb chronology
136–137
Chicago Pile number 1
65–70
at Hanford Works 77
invention of the nuclear
reactor 50–55
at Los Alamos 78
and secrecy 63, 81
and U.S. government
60–61
Fernald Feed Materials
Production Facility 139
Feynman, Richard 88, 90
final safety analysis report
(FSAR) 116
fire 1

FirstEnergy Corp. 140
fissile 132–133
fission *See* nuclear fission
fluorescence 11
Ford, Gerald R. 125, 138
Formerly Utilized Sites
Remedial Action Program
(FUSRAP) 138
fractional crystallization
42–43
France 122
F-Reactor 78
Frisch, Otto Robert 45, 46,
136
FSAR (final safety analysis
report) 116
Fuchs, Klaus E. J. 81–82
Fukushima I Nuclear Power
Plant 121, 130–131, 141

G

gadget 78 *See also*
Manhattan Project
gamma rays 19, 27, 39
gaseous diffusion 75
Geiger, Hans 23–25, 135
Geiger counters 52
General Electric 95
Germany
atomic bomb programs
48–50, 83–85, *84*
power plants 122
refugees 44, 50, 52–53,
55–56
global warming 130
Gold Foil Experiment 23–26,
24, 135
graphite moderation
Chernobyl RBMK 101,
129, 139
Chicago Pile number 1
64–70, *66, 67*
MAGNOX reactors *120*

and *Nautilus* design 95
Windscale reactor 101–
103, 138
Graves, A. C. 65
Great Britain *See* United
Kingdom
greenhouse gases 133
ground state 31
Groves, Leslie Richard 71, 74,
77, 83, 136

H

Hahn, Otto 41–44, 135
half-life 20, 27–28, 124
Hanford National Laboratory
105
Hanford Works (Site W) 73,
76, 93
Harteck, Paul 136
heavy water (deuterium-
oxide) moderation 49, 104,
119–120, *122*
Heisenberg, Werner 52
helium 20–21, 35–36, *37*
Hertz, Heinrich Rudolf 8, 134
high-voltage coils 8–9
Hilberry, Norm 66
Hirohito (emperor of Japan)
92
Hiroshima, Japan 90–92,
91m
Hitler, Adolf 44
Home Chain radar 48
Hooper, Stanford C. 60–61
hydrogen 30–31, 134
hydrogen bomb 55–56
hypercritical assembly 78–79,
80

I

implosion 80, 82–83, 93
inert gases 34

Institute for Theoretical
    Physics 135
Institute of Nuclear Power
    Operations (INPO) 139
interaction 52–54
interaction probability 54
International Atomic Energy
    Agency 137

J

Japan
    Fukushima I Nuclear
        Power Plant 121, 130–131,
        141
    Hiroshima and Nagasaki
        90–92, *91m*
    nuclear bomb program
        85
Jews and Jewish heritage
    44–45, 50, 52–53, 55–56
Joliot-Curie, Irène and
    Frédéric 37, 135

K

K-25 75–76
KANUPP reactor 124
Kristiakowsky, George 78,
    83
krypton 43
Kurchatov, Igor Vasilevich
    136

L

Lavoisier, Antoine-Laurent 4,
    4, 134
Lawrence, Ernest 57–59, *58*,
    70, 74, *75*, 136
Lenard, Philipp 21–23, 25–26
Leucippus 2, 134
Lichtenberger, Harold 65

light
    Balmer Series 30–31
    particle/wave nature of
        32–33
    Planck's constant 28–29
light water *See* water
    moderation
Lilienthal, David 104
line spectra 31, 134
Little Boy (Thin Man, gun-
    barrel design) 80, 83, *86*,
    86–87, 94, 137
Los Alamos Laboratory
    (Site Y)
    Bethe, Hans 55
    Bohr at 53
    chronology 136, 137
    deaths at 109
    neutron research 78–80
    role of 73
    Teller, Edward 55–56
    uranium enrichment
        74–76
Los Alamos National
    Laboratory 105

M

MAGNOX reactors 118–119,
    120
Manhattan Project 70–92
    after World War II 93–94
    chronology 136–137, 138
    declassification of
        information 105
    espionage 81–82
    Fat Man (implosion
        design) 80, 82–83, *87*,
        87–88, 92, 93, 137
    Hanford (Site W) 73
    Hiroshima and Nagasaki
        90–92, *91m*

Little Boy (Thin Man,
    gun-barrel design) 80,
    83, *86*, 86–87, 90–92,
    91m, 94, 137
Los Alamos Laboratory
    (Site Y) *See* Los Alamos
    Laboratory (Site Y)
neutron research at Los
    Alamos Laboratory
    78–80
Oak Ridge (Site X) 71, 73
people and locations
    70–72, 73–74
plutonium-239 (Path 2)
    76–78, 80, 82–83, *87*,
    87–88
Trinity test 88, *89*, 90
uranium enrichment
    (Path 1) 74–76
        K-25 sub-path 75–76
        Y-12 sub-path 74, *75*
Marsden, Ernest 23–25, 135
MAUD Committee 64, 70–71
Maxwell, James Clerk 7–8,
    18, 30
McMahon Act (1946) 104
Meitner, Lise 42–45, 136
Mendeleev, Dmitry 34
MeV (million electron volts)
    54
moderators *See* graphite
    moderation; heavy
    water (deuterium-oxide)
    moderation; water
    moderation
Monsanto Company 96
Murphree, Eger 70

N

Nagaoka, Hantaro 135
National Academy of Science
    125

national laboratories, U.S. 104, 111 *See also specific laboratories*

natural reactors 6

Nature 38, 51

Nautilus, USS 94–101, *98, 100,* 137

advantages of nuclear power 94

design requirements 95–97

effects on future nuclear power plant design 99–100

Nimitz approval of 97–98

safety 112–113

zirconium production 98–99

Navy, U.S. 60–61, 112–113 See also *Nautilus,* USS

Neddermeyer, Seth 78, 80, 82–83

neutrons 35–41, 52–54, 132–133 *See also* chain reactions, self-sustaining

New System of Chemical Philosophy (Dalton) 4–5

Nimitz, Chester 97–98

Nishina, Yoshio 85, 137

Nixon, Richard M. 125, 138

NRC (Nuclear Regulatory Commission) 125, 138, 140

NRG Energy, Inc. 140

NRX meltdown 104

nuclear accidents

Chernobyl 101, 129, 139

chronology 138–141

Fukushima I Nuclear Power Plant 121, 130–131, 141

NRX meltdown 104

Three Mile Island reactor number 2 (TMI-2) 128–129, 138, 139, 140

Windscale 101–103, 138

nuclear fission, discovery of 35–46, 136

Chadwick's proposed existence of neutrons 36–38, *38*

delayed, thermal fission 79

Fermi's nuclear reactor 51–55

Hahn and Meitner's fission of uranium 42–44

prompt, fast fission 79–80

Szilárd's self-sustaining chain reaction 38–41, *40*

uses of 44–46

nuclear fuel reprocessing 138, 139

nuclear physics 22

nuclear power plants/nuclear reactors

as application of fission 45

atomic bomb v. 79–80

boiling water reactors (BWRs) 107–108, 121, *123,* 137

BORAX experimental boiling water reactors 106–108, *107,* 137

Calder Hall 103, 116–117, *118,* 119, *120,* 138

countries claiming to be first 94

experimental v. uniform (standardized) designs 130, 139

first nuclear reactor (Chicago Pile 1) 35, 64–70, *66, 67,* 136

pressurized water reactors (PWRs) 97, 100, 106–107, 112–116, *113, 114,* 121

Sodium Reactor Experiment 111–112, 138

nuclear/radioactive waste 124–128, *126,* 138, 139, 140–141

Nuclear Regulatory Commission (NRC) 125, 138, 140

Nuclear Waste Policy Act (1982) 127, 139

Nuclear Waste Repository Act, PL 97-425 (1982) 125

nucleus *See* atomic nucleus, discovery of; nuclear fission, discovery of

Nyaya school 2, 134

Nyler, W. E. 65

O

Oak Ridge Laboratory (Site X) 71, 93, 95, 96, 137

Oak Ridge National Laboratory 105

Oklo uranium mine (Africa) 6

"On the Construction of a 'Superbomb' Based on a Nuclear Chain Reaction in Uranium" (Peierls) 46

Oppenheimer, J. Robert 73, 78, 88, 93, 136, 137

orbits 30–31

Organization of Petroleum Exporting Countries (OPEC) 138

Overbeck, Wilcox 70

P

Pakistan, KANUPP reactor
124
particle acceleration 57, 78
Peierls, Rudolph 46
periodic table of the
elements 33–34
photons 29
Pierrelatte Uranium
Enrichment Facility
(France) 6
pitchblende 17
Planck, Max 28–29, 135
Planck's constant 28–29, 135
planetary model of the atom
135
plum pudding model of the
atom 14, 21, *26*
plutonium-239 (Pu-239)
breeder reactors 106–108,
107
Manhattan Project
76–78, 80, 82–83, *87,*
87–88
plutonium-240 (Pu-240) 80
polonium 17
"Possible Existence of a
Neutron" (Chadwick) 38,
135
prehistoric nuclear activity
6
preliminary safety analysis
report (PSAR) 116–117
pressurized water reactors
(PWRs) 97, 100, 106–107,
112–116, *113, 114,* 121
Price-Anderson Act 138
prompt criticality 79
protons 27–28
PSAR (preliminary safety
analysis report) 116–117
PWRs (pressurized water
reactors) 97, 100, 106–107,
112–116, *113, 114,* 121

Q

Qinshan Phase III Nuclear
Power Units 1 and 2 *121*
quantum mechanics 27–34
Bohr's model of electron
orbit 30–32, *31*
particle/wave nature of
light 32–33, *34*
and periodic table 33–34
Planck's constant 28–29
threshold of 27–38

R

Rabi, Isador I. 78
radioactive decay 19–20, 21,
27–28, 57, 135
radioactive waste *See*
nuclear/radioactive waste
radioactivity, discovery of
12, 15–17, 134
radionuclides 124
radio waves 134
radium 17
Rambo, Sylvia 140
RBMK reactors 101, 121, 129,
139, 143
reactors *See* nuclear power
plants/nuclear reactors
Reagan, Ronald 139
Regulations for the Safe
Transport of Radioctive
Material 127
Resource Conservation and
Recovery Act (RCRA) 125,
138
Rickover, Hyman 95, 96,
97–100, 102–103, 112–114
Rocky Flats Plant 139
Roddis, Lou 95
Roentgen, Wilhelm 8–10, *9,*
10–11, 134
Rome–La Sapienza,
University of 51–54

Roosevelt, Franklin D. 62,
64, 136
Russia 122, 140 *See also*
Soviet Union
Rutherford, Ernest
alpha and beta rays
(particles) 19–21, *20*
chronology 135
discovery of the nucleus
22, 23–26, *24,* 35
on practicality of nuclear
power 39–40
Rydberg, Johannes 134
Rydberg constant 134

S

Sabath, Adolph 102–103
Sachs, Alexander 62, 64
safety *See also* Atomic
Energy Commission
(AEC); nuclear accidents
boiling water reactors
(BWRs) 107–108
experimental design and
130
final safety analysis report
(FSAR) 116
nuclear/radioactive waste
124–128, *126,* 138, 139,
140–141
preliminary safety
analysis report (PSAR)
116–117
RBMK reactors 101, 121
Sodium Reactor
Experiment core
overheating 111–112
U.S. Navy nuclear
submarine program 96,
97, 112–113
U.S. power plants 109–
110, 140

Salado and Castile Salt Formations 126–127

Santa Susana Field Laboratory (SSFL) 111

saturnian model of the atom 135

self-sustaining chain reactions *See* chain reactions, self-sustaining

Serber, Robert 78

Shippingport Atomic Power Station 114–116, *115,* 138, 139

silver 74

Site W *See* Hanford Works (Site W)

Site X *See* Oak Ridge Laboratory (Site X)

Site Y *See* Los Alamos Laboratory (Site Y)

Skeptical Chymist (Boyle) 3, 134

Soddy, Frederick 19–20, 135

Sodium Reactor Experiment 111–112, 138

Soviet Union
  AM-1 nuclear power station 101
  atomic bomb research 136
  Chernobyl accident 101, 129, 139
  espionage 63–64, 81–82
  nuclear research 85
  RBMK reactors 101, 121, 129, 139, 143

spectral lines 31, 134

Speer, Albert 84

spent-fuel shipping casks 127

spent-fuel storage 124–128, *126,* 138, 139

Stalin, Joseph 85

Steen, Charles A. 105

Strauss, Lewis L. 109–111

strong nuclear force 39, 50

strontium-90 124

"Studies in the Electron Theory of Metals" (Bohr) 30

submarines 103, 112–113, 117–118 *See also Nautilus,* USS

Superphoenix Nuclear Power Station 122

Suzuki, Kantaro 90

Szilárd, Leó
  Chicago Pile number 1 65, 70
  chronology 136–137
  Manhattan Project 78, 135
  neutronic reactor patent 50, 70
  nuclear chain reactions 22, 38–41, *40*
  and nuclear weaponry research 60, 62

T

Teller, Edward 55–56, 61, 62–63, 78, 136

thermal fission 79

thermal speed 45–46, 54

Thin Man 80 *See also* Little Boy (Thin Man, gun-barrel design)

Thomson, Joseph John "J. J." 12–15, *13,* 134

thorium 17, 19–20, 135

Three Mile Island reactor number 2 (TMI-2) 128–129, 138, 139, 140

Trinity (TR) 88, *89,* 90

Truman, Harry S. 90, 100, 104–105, 137

U

U-235 6, 46, 47, 74–76, *75,* 140

U-238 46, 47, 74–76, 77, 87, 136

Ulam, Stanislaw 78

United Kingdom
  Calder Hall nuclear plant 103, 116–117, *118,* 119, *120,* 138
  Home Chain radar 48
  MAUD Committee 64, 70–71
  spent fuel shipping 127
  submarine reactor 117–118
  Windscale reactors 101–103, 138

United States *See also* atomic bombs; Manhattan Project
  as center of nuclear research xii–xiii
  current state of nuclear power in xi–xii
  early nuclear research 56–58
  national laboratories 104, 111 *See also specific laboratories*
  reactors, numbers of 122–124, 140
  safety record 140
  standard v. experimental designs 130
  uranium supplies 94–95, 96, 105, 136

Untermyer, Samuel 106, 108

uranium 94–95 *See also* U-235; U-238
  the Curies and 15, 17
  isotopes 47 *See also* U-235; U-238
  U.S. supplies 94–95, 96, 105, 136

Uranium Club 48, 49

Uranium Committee 70, 136

uranium enrichment 74–76

uranium hexafluoride *75*

Uranium Mill Tailings Radiation Control Act 138

Urey, Harold 70

V

Vaisheshika school 2, 134

Van de Graaff linear particle accelerators 78

Venora Project 82

Villard, Paul 19

W

Waste Isolation Pilot Plant (WIPP) 125–127, *126,* 140

Waste Policy Amendment Act 139

water moderation

boiling water reactors (BSRs) 107–108, 121, *123,* 137

pressurized water reactors (PWRs) 97, 100, 106–107, 112–116, *113, 114,* 121

Watts Bar Nuclear Power Plant 140

weak nuclear force 50

Weil, George 65, 68, 69

Weizsäcker, Carl Friedrich von 48–49

Westinghouse Corporation 95, 98, 99

Wheeler, John A. 77–78

Wigner, Eugene "E. P." 56, 61, 62, 136

Wilkinson, Eugene P. 100

Windscale accident 101–103, 138

Woods, Leona 66–67

World War II *See* atomic bombs; Manhattan Project

X

xenon-135/xenon poisoning 77–78

X-rays *9,* 9–10

Y

Y-12 74, *75*

Young, Thomas 32

Yucca Mountain 127–128, 140–141

Z

Zaporizhzhia Nuclear Power Plant 124

Zinn, Walter 66, 68, 106

zirconium 98–99